MANNED LABORATORIES IN SPACE

Symposium Organized by the International Academy of Astronautics

ASTROPHYSICS AND SPACE SCIENCE LIBRARY

A SERIES OF BOOKS ON THE RECENT DEVELOPMENTS
OF SPACE SCIENCE AND OF GENERAL GEOPHYSICS AND ASTROPHYSICS
PUBLISHED IN CONNECTION WITH THE JOURNAL
SPACE SCIENCE REVIEWS

VOLUME 16

MANNED LABORATORIES IN SPACE

Second International Orbital Laboratory Symposium

Organized by the

INTERNATIONAL ACADEMY OF ASTRONAUTICS

at the XIXth International Astronautical Congress

New York, 18 October 1968

Edited by

S. FRED SINGER

D. REIDEL PUBLISHING COMPANY

DORDRECHT-HOLLAND

Printed in The Netherlands by D. Reidel, Dordrecht

The low-salinity wedge and the coast of Georgia. The Atlantic Ocean and the coast of Georgia taken during the flight of the Apollo 502 unmanned spacecraft, April 4, 1968, with a 70-mm focal length lens. (NASA color photograph AS6-2-1485.) See also the paper by Wenk, below, pp. 68–74.

PREFACE

The publication of this book is extremely timely, for the next major advances in manned space flight after Project Apollo will most likely be made in earth orbital operations.

Manned exploration of the moon will certainly continue after the initial landing, but it will be performed essentially with the Saturn V launch vehicles and Apollo spacecraft developed in Apollo, especially in the early phases. Modifications to this basic hardware will increase operating capabilities to permit extensive lunar exploration during prolonged staytimes by the astronauts on the moon's surface.

Manned orbital space stations have been studied for years, and NASA is already well along in development of its first attempt to provide more spacious accommodations for astronaut-scientists in its Saturn Workshop program.

While the Workshop is certainly not the ultimate space station of which our technology is capable, it is a workable, poor man's approach to the immediate need for using and expanding our present manned space flight capability without a detrimental loss of momentum. The approach of converting a Saturn rocket stage into a manned laboratory and observatory in space is an improvisation that matches the use of the jerry-built Jupiter C back in 1958 to launch Explorer I. Let's hope that it can get the job done just as effectively.

I believe that the Workshop will prove valuable as a research tool for increasing our knowledge of the effective integration of man into the total system. The results should help determine the systems configuration and operational approach for true space stations such as those envisioned in this volume.

WERNHER VON BRAUN

EDITOR'S PROLOGUE

As I write this prologue, we stand at the threshold of man's first step onto another celestial body. As you read these words, this historic step will have been taken. The word 'historic' may sound a little trite perhaps, but it will be considered as a historic event a hundred years from now, a thousand years from now, and for as long as the history of mankind is written.

Manned spaceflight began with Yuri Gagarin's orbit around the earth in Vostok 1 in April 1961. Many other flights have followed, with increasing complexity and increasing number of tasks for the occupants (see Table I). The concept of a manned laboratory in space has become firmly established as a logical outgrowth of these pioneering flights.

Almost at the eve of the manned lunar landing, on October 18, 1968, the International Academy of Astronautics arranged a Symposium on Manned Space Laboratories in New York City. As program chairman, I commissioned and arranged the lectures and added three further papers to make up this volume. We start with an overview of manned space programs by Dr. Robert Gilruth, Director of the Manned Spacecraft Center of the National Aeronautics and Space Administration in Houston, Texas. The following papers fall into four major areas: First, the various applications of a manned orbiting laboratory for observations of the earth and its terrain, of the oceans, and of the atmosphere. Next follows a discussion of the manned astronomical laboratory for the observations of objects beyond the earth.

The third area is technologically oriented and discusses the design problems of a manned space station in a comprehensive fashion. The final area concerns the use of a manned space station as a stepping stone for manned planetary flight, with a particular case being made to adopt a flyby of Mars and Venus as a next major space goal.

The papers in the volume are addressed to the general reader, but contain a great deal which will be of interest to the technological specialist, as well as to the geophysicist and astronomer. The volume presents a 'snapshot' as it were, of the thinking which prevailed just before man set foot on the moon. It represents a record and projection, an assessment and a prediction, taken at the beginning of the second decade of the space age. Some years from now, when we look back on the thoughts expressed by the contributors to this volume, we will be able to judge whether they were visionary or conservative, fantastic or realistic.

S. FRED SINGER

Washington, D.C., June 1969

TABLE I

Name	Int'l desig.	Proj. dir.	Launch data			Initial orbital data					Status
			Date	Site	Vehicle	Weight	Period	Perigee	Apogee	Incl.	
Vostok 1	1961 M1	USSR	4/12/61	Tyuratam	A-1	10417	89.1	112	203	65.0	Re-entered 4/12/61; first manned space flight, cabin with Y. Gagarin recovered in USSR after 1 orbit, 1.8 hours.
Vostok 2	1961 T1	USSR	8/6/61	Tyuratam	A-1	10430	88.6	111	160	64.9	Re-entered 8/7/61; G. Titov landed after 17 orbits, 25.3 hours.
Mercury-Atlas 4	1961 AA1	NASA	9/13/61	ETR	Atlas D	2700	88.6	100	159	32.6	Re-entered 9/13/61; recovered from ocean after 1 orbit.
Mercury-Atlas 6	1962 1	NASA	2/20/62	ETR	Atlas D	2987	88.5	100	163	32.5	Re-entered 2/20/62; first U.S. manned orbital mission, J. Glenn and 'Friendship 7' recovered after 3 orbits, 4.9 hours.
Mercury-Atlas 7	1962 T1	NASA	5/24/62	ETR	Atlas D	2975	88.3	100	167	32.5	Re-entered 5/24/62; 'Aurora 7' and S. Carpenter recovered after 3 orbits, 4.9 hours; landed 250 miles from target.
Vostok 3	1962 AM1	USSR	8/11/62	Tyuratam	A-1	10410	88.5	114	156	65	Re-entered 8/15/62; A. Nikolayev landed by parachute after 64 orbits, 94.4 hours; part of first Soviet 'group' flight.
Vostok 4	1962 AN1	USSR	8/12/62	Tyuratam	A-1	10423	88.5	112	158	65	Re-entered 8/15/62; P. Popovich landed by parachute after 48 orbits, 71.0 hours; came within 3.1 miles of Vostok 3 on orbit 1.
Mercury-Atlas 8	1962 B 1	NASA	10/3/62	ETR	Atlas D	3029	89.0	100	176	32.5	Re-entered 10/3/62; 'Sigma 7' with W. Schirra recovered within 5 miles of carrier after 6 orbits, 9.2 hours.
Mercury-Atlas 9	1963 15A	NASA	5/15/63	ETR	Atlas D	3033	88.5	100	166	32.5	Re-entered 5/16/63; 'Faith 7' with L. G. Cooper recovered after 22 orbits, 34.3 hours; completed Mercury program.

Table I (continued)

Name	Int'l desig.	Proj. dir.	Launch data			Initial orbital data					Status
			Date	Site	Vehicle	Weight	Period	Perigee	Apogee	Incl.	
Vostok 5	1963 20A	USSR	6/14/63	Tyuratam	A-1	10406	88.4	112	146	65	Re-entered 6/19/63; V. Bykovsky landed by parachute after 81 orbits, 119.1 hours; part of second 'group' flight.
Vostok 6	1963 23A	USSR	6/16/63	Tyuratam	A-1	10390	88.3	114	145	65	Re-entered 6/19/63; V. Tereshkova landed by parachute after 48 orbits, 70.8 hours; passed within 3 miles of Vostok 5.
Voskhod 1	1964 65A	USSR	10/12/64	Tyuratam	A-2	11728	90.1	111	254	65	Re-entered 10/13/64; first three-man crew – V. Komarov, K. Feokistov, B. Yegorov; landed after 16 orbits, 24.3 hours.
Voskhod 2	1965 22A	USSR	3/18/65	Tyuratam	A-2	12527	90.9	107	308	65	Re-entered 3/19/65; A. Leonov spent 20 minutes outside spacecraft, landed with P. Belyayev after 17 orbits, 26.0 hours.
Gemini 3	1965 24A	NASA	3/23/65	ETR	Titan II	7111	88.2	100	140	32.5	Re-entered 3/23/65; first manned orbital maneuvers, V. Grissom and J. Young landed after 3 orbits, 4.9 hours.
Gemini 4	1965 43A	NASA	6/3/65	ETR	Titan II	7879	89.0	100	175	32.0	Re-entered 6/7/65; E. White performed 21-minute EVA experiment, landed with J. McDivitt after 66 orbits, 97.9 hours.
Gemini 5	1965 68A	NASA	8/21/65	ETR	Titan II	7947	89.5	104	195	32.6	Re-entered 8/29/65; first extended U.S. manned flight, L. G. Cooper and C. Conrad landed after 128 orbits, 190.9 hours.
Gemini 7	1965 100A	NASA	12/4/65	ETR	Titan II	8076	89.2	100	204	28.9	Re-entered 12/18/65; F. Borman and J. Lovell went record 220 orbits, 330.6 hours; served as Gemini 6 rendezvous target.

Table I (continued)

Name	Int'l desig.	Proj. dir.	Launch data		Vehicle	Initial orbital data					Status
			Date	Site		Weight	Period	Perigee	Apogee	Incl.	
Gemini 6	1965 194A	NASA	12/15/65	ETR	Titan II	7817	88.5	100	162	28.9	Re-entered 12/16/65; W. Schirra and T. Stafford rendezvoused within 1 foot of Gemini 7, landed after 17 orbits, 25.9 hours.
Gemini 9	1966 47A	NASA	6/3/66	ETR	Titan II	8268	88.8	100	169	28.9	Re-entered 6/6/66; rendezvous and EVA tests carried out by T. Stafford and E. Cernan; landed after 47 orbits, 72.3 hours.
Gemini 10	1966 66A	NASA	7/18/66	ETR	Titan II	8295	88.8	100	168	28.9	Re-entered 7/21/66; rendezvoused with Gemini 8 and 10 targets; J. Young and M. Collins landed after 46 orbits, 70.8 hours.
Gemini 11	1966 81A	NASA	9/12/66	ETR	Titan II	8374	89.0	100	174	28.8	Re-entered 9/15/66; C. Conrad and R. Gordon performed initial first orbit docking; landed after 47 orbits, 71.3 hours.
Gemini 12	1966 104A	NASA	11/11/66	ETR	Titan II	8294	89.0	100	175	28.9	Re-entered 11/15/66; J. Lovell, E. Aldrin ended program with successful EVA tests; landed after 63 orbits, 94.6 hours.
Soyuz 1	1967 37A	USSR	4/23/68	Tyuratam	A-2		88.6	125	139	51.7	Re-entered 4/14/67; recovery attempt after 18 orbits, 26.8 hours; failed due to fouled parachutes; V. Komarov killed.
Apollo 7	1968 89A	NASA	10/11/68	ETR	Saturn 1B	45089	89.7	142	177	31.64	Re-entered 10/22/68; astronauts Schirra, Cunningham, and Eisele complete near-flawless first manned flight of Apollo spacecraft; recovered after 163 orbits, 260.2 hours.
Soyuz 3	1968 94A	USSR	10/26/68	Tyuratam	A-2		88.6	127	140	51.4	Re-entered 10/30/68; Cosmonaut Beregovoy maneuvered to 650 feet from Soyuz 2; landed after 64 orbits, 94.9 hours.

Table I (continued)

Name	Int'l desig.	Proj. dir.	Launch data		Vehicle	Initial orbital data					Status
			Date	Site		Weight	Period	Perigee	Apogee	Incl.	
Apollo 8	1968 118A	NASA	12/21/68	ETR	Saturn V	63650	88.15	114	118	32.6	Re-entered 12/27/68; first manned circumlunar flight; astronauts Borman, Lovell and Anders recovered in mid Pacific after 10 lunar orbits, 147 hours.
Soyuz 4	1969 4A	USSR	1/14/69	Tyuratam	A-2	16,000	88.25	107	140	51.40	Re-entered 1/17/69: Kosmonaut Shatalov performed manual docking with Soyuz 5 in first link-up of two manned vehicles; landed on 48th orbit after 71.2 hours.
Soyuz 5	1969 5A	USSR	1/15/69	Tyuratam	A-2	16,000	88.7	124	143	51.40	Re-entered 1/18/69: Kosmonauts Khrunov, Yeliseyev performed 1 hour EVA, transferred to Soyuz 4 in rehearsal of emergency rescue; Kosmonaut Volynov landed alone on 49th orbit after 72.8 hours.
Apollo 9	1969 18A	NASA	3/3/69	ETR	Saturn V	90,585		118	120		Re-entered 3/13/69: McDivitt, Scott and Schweickart perform successful first manned flight of Apollo LM; splash down after 151 orbits, 241.9 hours, including 40 minutes EVA. Ascent stage decayed 3/22/69.
Apollo 10	1969 43A	NASA	5/18/69	ETR	Saturn V	94,497	88.1	115	115	32.5	Re-entered 5/26/69: Second manned circumlunar flight; LM descended to within 9 miles of lunar surface; Astronauts Stafford, Young and Cernan recovered Mid Pacific after 192 hours.

TABLE OF CONTENTS

MANNED SPACE STATIONS – GATEWAY TO OUR FUTURE IN SPACE

R. R. GILRUTH

NASA Manned Spacecraft Center, Houston, Tex., U.S.A.

Abstract. The paper describes several manned space station concepts ranging from early versions, which will provide the basic understanding of man's ability to handle the space environment and to work in it productively, to the more complex space bases, which will cluster a large variety of specialized scientific satellites around a central hotel core. The rationale for choosing the size of the space station, the number and characteristics of its crew, the altitude and inclination of its orbit, and lifetime and resupply requirements are presented. The contribution of space stations toward fully utilizing man's unique reasoning and manipulative characteristics in the observation of the earth and the telescopic exploration of the universe is discussed, along with the role of space station development and operational experience in the creation of technology required to send man to the planets.

1. Introduction

Space stations orbiting the earth have long been the dream of space engineers and scientists. There have been many concepts and designs proposed, and many reasons have been given for their use. One of the problems has been the difficulty of dealing with the very multitude of uses and the infinite variations of designs.

During the last few years, however, several things have happened to help clarify the picture: First, the recent development of the Saturn V rocket has provided one of the major building blocks for space station design.

Secondly, as a result of our manned flight experience we now have a far greater knowledge of the environment desirable for working and living in space; and,

Thirdly, the programs of the work in science and technology to be done in a space station have been studied in considerable depth with the result that the use requirements can now be better defined.

2. Why Space Stations?

A space station is not so much a 'thing' as it is a location in space which is developed to support men and equipment on a permanent basis in order to take advantage of the economies of size, centralization, and permanency. In other words, a space station is a 'base' in space, equivalent in function to those used in many forms of terrestrial exploration; for example, those in Antarctic. The base has six important characteristics: it is a central location for power, enclosed volume, logistics, experiment, equipment, communications, and data systems. Such a definition encompasses schemes in which all of these things might be located within a single volume, or alternately where there may be several devices which are orbiting the earth in close proximity with one another. An example of this latter form would be a station which

S. F Singer (ed.), Manned Laboratories in Space.

consists of a large living center surrounded by free modules containing telescopes, earth sensors, and specialized scientific satellites orbiting close enough so that they could be tended or serviced intermittently by personnel from the living quarters.

We are seeking the economic benefits which will accrue from the longer lifetime of the objects we place in space. Both the equipment required to support man and to run the station, and the basic scientific equipment which we place in space can operate for long lifetimes. With proper logistics support, these items can be re-used, can be refurbished where wearout phenomena or random failures have occurred, and, perhaps more importantly, can be modified to take advantage of gains in technology or changes in experiment objectives.

As regards *logistics*, the creation of a permanent place in space to which we return again and again allows us to centralize the storage of expendables and equipment and to make maximum use of a limited number of logistics flights. Since logistics flights are the largest single cost element in any orbital program, this is an extremely important advantage of space station operation.

Now as regards *power systems*, the large size of the space station enables us to develop the structure necessary to support either the large panels required for solar cells or to provide the separation, shielding weight, and radiator areas required for large nuclear power systems. The capability of a space station to carry power sources of 20 to 50 or even 100 kilowatts will be a most important technical contribution, since the availability of large amounts of power is fundamental to the operation of nearly all experiment equipment and will enhance the standard of living in space.

There are important advantages to the creation of a so-called personnel center in space. First, when we have achieved a crew size of over 10 to 12 people, we begin to realize the economy of large size since a progressively smaller percentage of total crew worktime will have to be spent on the functions of operating and maintaining the station, thus making available more time for space experiments. The economies of specialization are also available as the concentration of larger number of people allows more of them to be specialized for individual functions. This would relieve the considerable cross-training problems which are inherent in small crews. The concentration of personnel in a permanently established volume makes it feasible to provide relatively large amounts of volume devoted to personal comfort and convenience, and to establish, throughout the station, living conditions which are more earth-like than we can afford in smaller space vehicles. This will be important in the selection and training of crews and in their efficiency in space operations as well.

3. Post-Apollo Program (Apollo Applications) – A Modest Beginning

We have long recognized the intrinsic advantages of space stations and the Post-Apollo Program is designed to explore them in a modest way. The primary objective of this program is the establishment of man's capability to function in space for significantly long periods of time and to establish man's role in conducting complex investigations in space. The orbital workshop, for example, will contain sufficient volume (almost 4 000 cubic feet per man) and support equipment so that the astronauts will be able to function and live in a manner more like that of the earth's environment than has been the case with the closely packed spacecraft to date. This will be our first experience with large amounts of enclosed volume in space. It is planned to conduct detailed medical measurements and also complex scientific experiments

The workshop will include a major science experiment, a solar observation experiment which will be launched unmanned and which, after the rendezvous, will be docked by an astronaut located in the docking adapter, using remote control. Figure 1

Fig. 1. Orbital workshop with solar experiment attached.

shows a view of this experiment located on the workshop. This experiment will provide us with two very important steps toward our future space station programs. First, it will be our first chance to link man with a really complex scientific experiment and thus to explore his capabilities to function with such devices. Secondly, we will be developing techniques and equipment for sending experiments and supplies to a space station in an unmanned mode. This mode may be an important method for logistic supply in the future.

It is apparent from this brief description that the workshop will be fulfilling the role of a prototype space station. It will provide a generous living volume, the essential power and communications systems, and it will be operated over a period of time. Thus, our 'modest' beginning will be clearly laying the groundwork for the development of future space stations by giving us basic data on man, the systems, and some of the experiments which will define our future programs.

In creating the workshop, however, we have not been able to include what may be one of the major requirements for successful operation of a space station – artificial gravity.

4. The Need for Artificial Gravity

Our concern is not that man will not be able to adapt to zero gravity over the period of time that he is likely to be a crew member aboard a space station (3 to 6 months). The problem is one of providing relatively normal living and working conditions. This will be important for crew comfort and adaptation to living conditions and will, therefore, increase their basic efficiency and task effectiveness.

The environmental factors affected by artificial *g* can be divided into three types: those connected with processes involving fluids; those connected with locomotion and orientation; and those including general man/machine interfaces. By providing artificial gravity, fluid processes such as associated with personal hygiene, cleaning, food preparation, chemistry, all can be performed in a manner identical to that which we are accustomed to here on earth. With the establishment of normal gravity relationships, the ability to walk with the hands free will provide the basis for an environment which needs little or no training for adaptation. Artificial gravity will also provide normal man/machine interfaces with all types of equipment – both operational and scientific. Again, this would eliminate the need for special zero-*g* training and would maximize the effectiveness of earth training, especially for the more complex repair, refurbishment, and modification tasks we can expect in the future. This advantage is a major feature of an artificial gravity station. As a corollary to this,

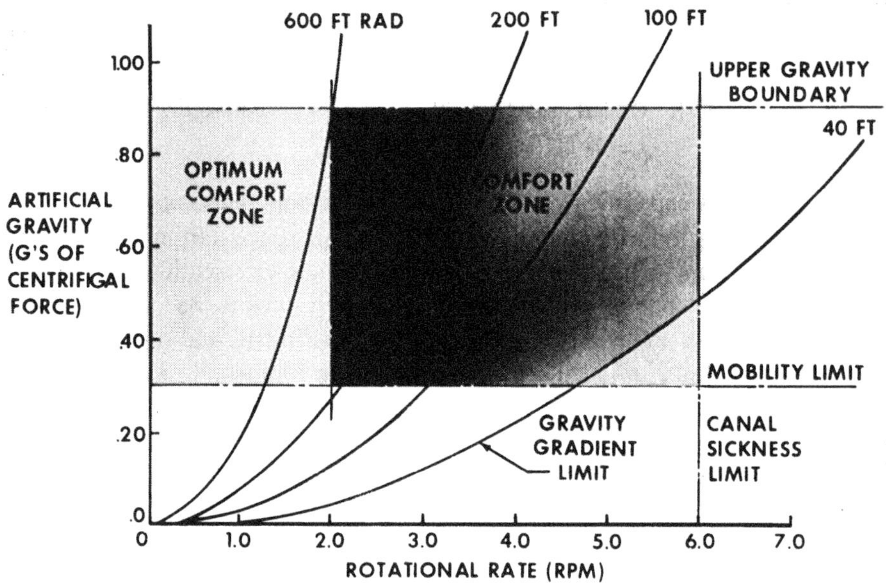

Fig. 2. Artificial gravity envelope.

artificial g will permit the use of equipment developed for use in earth laboratories but otherwise applicable to space station tasks.

For all of these reasons, we conclude that the characteristics of artificial gravity are important objectives for a near-term earth orbital activity. Some of the parameters which should be studied by special experiments are shown in Figure 2.

Artificial gravity has two important parameters for any given level: the rate at which the vehicle is rotated, and the radius at which the man is stationed. The 'comfort zone' is an area which ground-based tests have indicated man can tolerate without serious problems of adjustment. Parabolic airplane flights have indicated that most of the problems of locomotion and fluid transfer are overcome by gravity as low as three-tenths g. Since rotational simulators on the ground have indicated that the average person can adjust to 4 revolutions per minute, we are led to a minimum rotational arm of about 50 feet. However, these ground tests are limited because they always have a 1-g field affecting the results, and we cannot be certain that when that field is non-existent that the rotational forces will not affect the subconscious adaptation mechanisms more strongly. Thus we are interested in even lower rotational rates and higher gravity forces as high as 1-g. These considerations have led us to examine a system shown in Figure 3; namely, the rotation of an Apollo spacecraft with an experiment module, both tethered to a spent booster stage in such a way that various radii can be achieved by the system. With this type of experience, we can be ready to construct the type of space accommodations which we feel will be necessary to meet the comfort and work requirements of the future space activity.

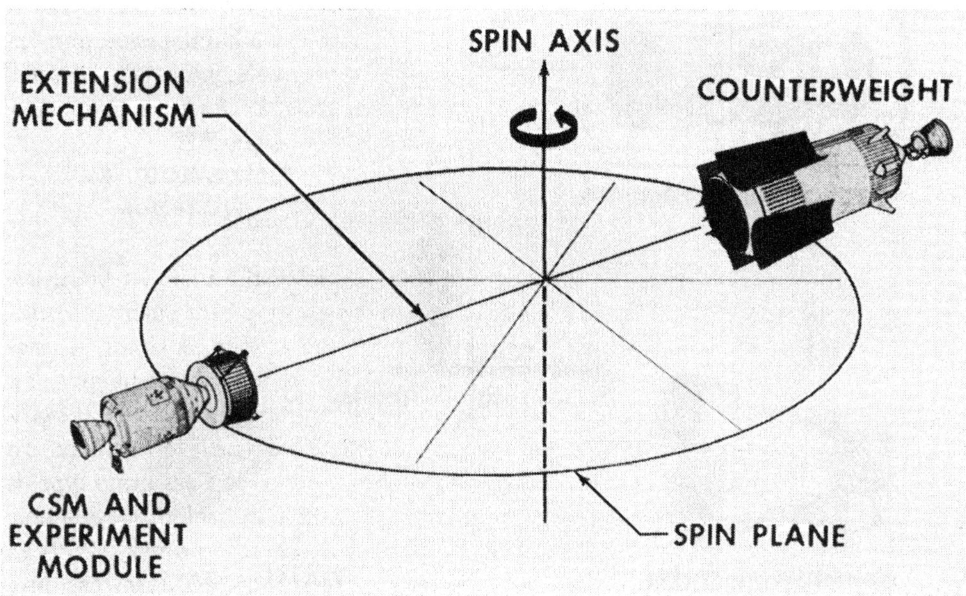

Fig. 3. Artificial gravity experiment.

5. The Next Step – Experiments and Applications

Activities to be conducted on board the space station of the future can be classified under two general headings: observational experiments which include astronomy and earth sensing; and onboard experiments such as bioscience and biomedical experiments and high energy physics.

6. Astronomy

The sciences which make use of the observational advantages of earth orbit will benefit most obviously from space station operations. Astronomy especially can be expected to produce dramatic results from being able to observe outside of earth's dirty and shimmering atmosphere.

Figure 4 presents a feeling for some of the engineering work that has gone into the sizing and design of astronomy experiments, such as modestly-sized telescopes of the 40-inch objective type, which are the optical equivalent of land-based 120-inch telescopes. Even these are large enough to require a significantly large space station as a base; but these are just a beginning. Telescopes of 120-inch diameter have been seriously proposed by the National Academy of Sciences as an objective for space

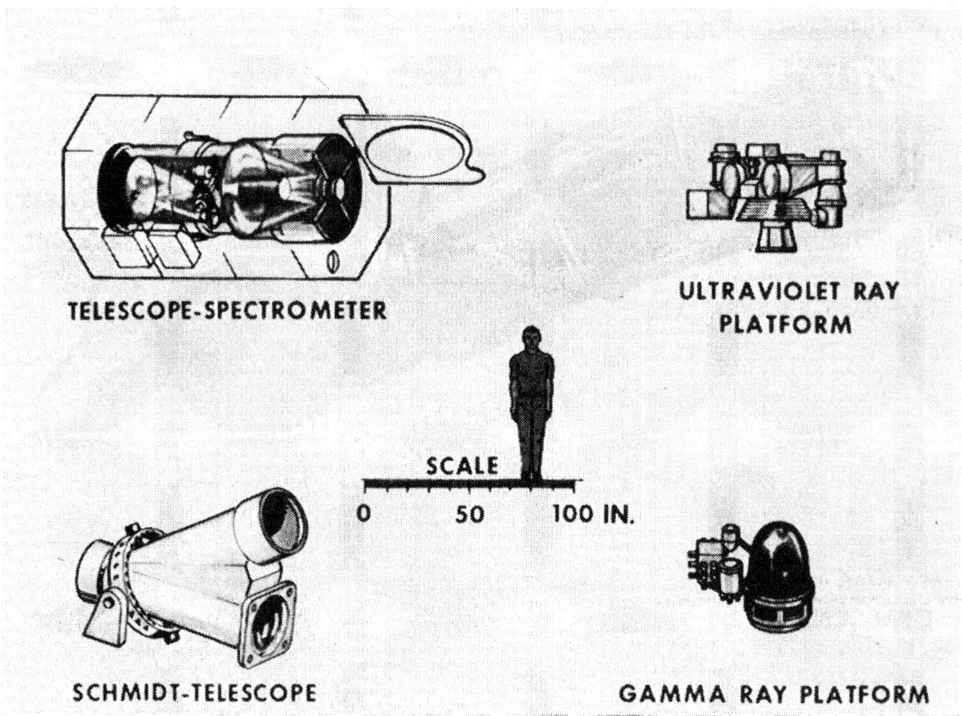

Fig. 4. Astronomical instruments for space.

astronomy. A conceptual design of a 120-inch telescope is shown in the background at the same scale to give a feel for the size of things to come. (See Paper 'Astronomical Research with a Large Orbiting Telescope', by Lyman Spitzer, Jr., p. 88.) Such a telescope would probably be flown adjacent to a space station and visited periodically for service and data retrieval.

7. Earth Sensing

The development of remote sensing equipment for the observation of the earth's surface, and probably subsurface, will be one of the major tasks of the manned space base.

To date, our principal experiments with earth observation from space have been primarily photographic. As many investigators have shown, these photographs taken with simple equipment give us dramatically new and astonishingly useful views of our home planet.

Evidence of the attributes and value of space photography are vividly shown in the Paper 'Earth Resource Observations from an Orbiting Spacecraft', by W. T. Pecora, p. 75.

The unique capabilities of space photography result from the broad synoptic view of large features which can be seen in a single photograph. Large mountain systems can be visualized in a single photograph, and important geophysical relationships are established over areas of up to 250 miles square.

Another use for space observation is oceanography. Techniques are being developed for routine observation of the ocean; the measuring of its temperatures, sea state, and currents. In this regard, see Paper 'Spacecraft for Ocean Exploration', by Edward Wenk, Jr., p. 68.

But photography covers only a very small part of the information the earth is sending into space. Electromagnetic earth sensing equipment has the capability of measuring the earth's surface over nearly the entire electromagnetic spectrum. The large variety of equipment which will be of interest in this area of investigation can easily be imagined. (See Paper 'The Research Potential of Manned Earth Orbiting Spacecraft for Meteorology', by S. Fred Singer, p. 33; and 'Terrain Spectral Imagery from Space in Various Spectral Regions', by K. Ya. Kondratyev *et al.*, p. 46.)

There is little doubt that this field of endeavor holds great promise for future applications of direct benefit to a wide variety of professions and occupations, as well as to the scientific community in general. It will probably be so useful that we will want to spread its benefits to other nations. This use of our space abilities may well become a great force for international cooperation and I look forward to the day when our space station crews will contain representatives from other nations of the world as well as our own.

8. Biomedical and Bioscience Experiments

Onboard experiments can take a great many forms. Some will exploit the limitless vacuum of space; these may well involve manufacturing processes. Some will be in space to explore the unique environment of zero gravity; biomedicine and bioscience will undoubtedly open new avenues in research on the growth process in animals and plants.

9. Physics

It is interesting to recall that the study of nature's fundamental particles began with observations of cosmic radiation, made from balloons and mountain tops. Space station capabilities reopen the natural environment as a primary source of high-energy particles. By placing a large cosmic-ray facility in a space station, we can extend the basic particle energies available for investigation by three orders of magnitude – from 70 to 70 000 Gev. Cryogenically-cooled superconducting magnets would form the heart of the facility. Through such a facility we might well expect new clues to such question marks as antimatter in the cosmos. Operating such a laboratory will be a very complex affair, involving the most advanced techniques for recording and analyzing data. Its character dictates the presence of scientists, not just astronauts.

10. The Next Step – Space Station Characteristics

Having examined the types and general characteristics of possible experiments, we can now proceed to examine what one of the first operational space stations might be like.

The first requirement is that the station have an earth-like environment. This means, of course, that the station will be designed to provide air at 10–14 psi and for one-half to 1-g of artificial gravity in the major living and operational areas of the station. A counter-rotating laboratory operating at essentially zero gravity is required for specialized scientific studies. The station should have the capability of accommodating at least a 50-man crew initially, with the prospect of being able to expand its capabilities to at least 100 by additional construction. We must certainly plan to operate the station at higher inclination than those conducted to date in our manned space flights. Our studies have shown that about 50° inclination, or somewhat higher, is a good compromise for operational and observational reasons. That inclination provides a near minimum of radiation background which is important for minimizing its effects on the crew and on the sensitive film used in astronomical observations. It is sufficiently high to cover the majority of the occupied land masses even though it leaves out some of the more developed countries in Europe. It would include, of course, all of the U.S.A. except Alaska, and it is within the continental U.S.A. that the majority of ground sites will be located which will be used to test orbital observation techniques.

The final general characteristic noted is that the station will have both attached and station-keeping experiment modules. For example, in the astronomical area we can expect attached telescopes up to 40 inches in diameter, while the proposed 120-inch telescope would be mounted in a separate module flying near the station in a station-keeping mode. This will mean the development of a technique for visiting such a

Fig. 5. Artist's concept of one million pound space station.

device for adjustments, data-gathering, and repairs. Smaller experiment modules will also probably be clustered around the station, and a hangar will be provided so that these modules can be brought back into the station for servicing by the crew.

Some details of this space station are shown in an artist's drawing, Figure 5. It would be large – having a 240-foot radius to the extremity of living quarters, and as much as a 375-foot radius to the balance and power section on the other end. Launched in three parts by three Saturn V rockets, the station would be assembled in space. It would have a gross weight of about one million pounds and would carry as much as 100 000 pounds of experimental equipment. Its living quarters would offer a volume of about 50 000 cubic feet. A zero-g laboratory at the hub could include 45 000 cubic feet of working space. The large radius would permit nearly 1-g at 3.5 rpm (thus maximum rotation needed). Having a rather large inertia, this station would require 7 000 pounds of propellant just for the spin-up to 3.5 rpm. Then it would have sufficient momentum to remain essentially fixed in inertial space.

Also shown is the 120-inch telescope in operation nearby the space station. Note the hangar, with a small satellite being taken in for maintenance. Small telescopes adjoin the zero-g hub on the top surface. Onboard experiments would be located both in the hub and in the artificial g volume. The hub would accommodate basic biomedical and bioscience zero-*g* experiments. The majority of other onboard experiments would go in the rotating-*g* section. That section would include complete machine shops and repair rooms, as well as the living quarters and command and data reduction areas.

It is important that such a station have growth possibilities. An additional launch of the Saturn V could, in fact, double the accommodations by delivering additional living space to the inhabited part of the station and balancing the other end by the booster which brought it up. A similar operation could add capability to the zero-*g* hub. The S-II stage, part of the station's basic structure, presents an even more economical way to expand living and working space. The empty tanks of S-II stages would contain over 36 000 cubic feet of volume that could be made serviceable using similar techniques we are developing with our Post-Apollo workshop.

I have tried to indicate to you the place of the space station in future space programs; the reasons why such stations will be important links with the future; the prototype steps we are taking ino ur present programs: and finally, a possible picture of the shape of things to come in terms of the conceptual design of a relatively large space station.

Such a station merges the space technology of the period with elements of both pure science and practical science into a single facility which can be logistically operated at a single place in space. Its purpose will be to thrust basic research into a new and stimulating environment. Permanency and vastness will be the station's predominant assets and it will provide a stately reservoir of energy and utilizable volume.

A philosophy of continuous re-use and the capability to expand its resources will enhance its long-term utility.

CRITICAL ENGINEERING PROBLEMS OF SPACE STATIONS

T. J. GORDON* and R. L. GERVAIS

Advance Space Stations and Planetary Systems,
McDonnell-Douglas Astronautics Company, Huntington Beach, Calif., U.S.A.

Abstract. From today's vantage point, the design of long-duration operational space stations appears feasible and practical. Surprisingly, only a few critical engineering problems exist for which no unique solutions exist today.

Long subsystem operating life can result from an optimum mix of redundancy, maintainability, and checkout; a major remaining problem is the determination of means of accelerating component life testing.

Of the methods available for coping with space station abort emergencies, only crew escape modules appear to provide the required short response time and high reliability.

That man can survive in a zero g environment is now proven fact; yet undetermined are the various psychophysiological therapeutic and remedial techniques needed to ensure man's effectiveness during long-duration missions.

Problems of lesser magnitude include certain important engineering developments which must be achieved to ensure that space stations will perform with the highest efficiency. These efficiency-improving issues include the subsystem optimization and interface matching of the stability and control, environmental control and life support, and power subsystems.

While several potential engineering problems of the long duration space stations are discussed, the absence of a large number of major problems is really significant.

1. Introduction

Before long-duration manned stations can be flown, system concepts must be translated into engineering designs and hardware. From today's engineering vantage point, these system concepts appear feasible and rather surprisingly perhaps, these programs involve only a few critical engineering problems. 'Critical engineering problems' are defined as those for which no unique solution exists today but which must be solved to implement an efficient, manned, long-duration space station with low risk. Within the scope of this operational definition, there appear to be three major remaining engineering issues:

(1) Attainment and demonstration of required subsystem operating life.

(2) Emergency crew return.

(3) Man's performance at zero g.

In addition, certain important engineering developments must be achieved to ensure that the space station performs with highest efficiency. Although these are not critical in the sense defined earlier, special attention is required to ensure that the space station can efficiently perform the job for which it is intended. These efficiency-improving issues include the subsystem optimization and interface matching of the (1) stability and control subsystem, (2) the environmental control and life support subsystem, and (3) the power subsystem. The critical problems, some of their potential

* Now Senior Research Fellow, Institute for the Future, Middletown, Conn., U.S.A

solutions, and the considerations involved in improving the efficiency and interface balance between the major subsystems will be discussed in this paper.

2. System Effectiveness

A delicate balance must exist between factors of maintainability, reliability, logistics, and checkout. Maintainability implies concepts of subsystem design and installation which permit routine servicing and repair by the crewmen in orbit. Reliability includes the expectation of proper subsystem functioning during the time of operation. High overall reliability levels can best be achieved through preflight qualification testing and redesign and/or through the design of redundant or self-repairing subsystems. Checkout capacity can range from simple hand-held meters to completely automatic systems to monitor the continued health of the space station performance. These factors are interrelated in a complex way.

A review of the effect of these concepts on space station layout, crew skills, logistics requirements, and subsystems design indicates the possibility of defining an optimum mix leading to definition of a system requiring minimum crew skills and cost; however, such an operations research solution is not available for the general problem.

Figure 1 shows a typical history of component failure rate as a function of time. In the early phases of component operation, high failure rates are experienced as a result of 'burn-in' failures. After these burn-in failures are excluded, a family of components usually exhibits a more constant failure rate; that is, the loss of components of a family per unit time remains constant. Towards the end of the expected life of the component a 'wearout' phase is experienced.

In a long-duration space station, marginal components can be excluded by an initial burn-in period. The subsystem design must be predicated on the anticipation of the constant failure rate of the family of components. If the anticipated operating failure rate is higher than required, redundancy must be included in the basic design.

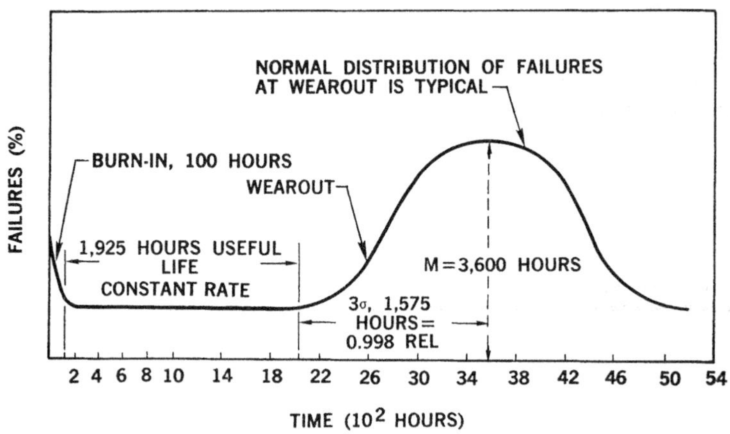

Fig. 1. Component failure, vacuum pump, diaphragm ECLS subsystem.

TABLE I

Typical 1-year space-station reliability/spares summary

Operational subsystem	Nonredundant basic reliability	Redundant without spares	With spares	Spares wt per 90 days (lb)
Electric power-solar cells with batteries	0.711	0.9918	0.9997	4.5
EC/LS	0.0276	0.0605	0.9901	287.4
Stability and control system	0.410	0.5125	0.9996	18.0
Communications and data management	0.0600	0.0819	0.9898	115.6
Reaction control/system	0.905	0.9954	0.9987	7.6
Structure	0.723	0.9561	0.9967	27.0
Summation, Σ	0.0003	0.0023	0.9747	443.9

The distribution of the wearout failures must be known adequately to determine the maintenance or replacement cycle for the subsystems. However, wearout distribution is difficult to determine for many components simply because operating lifetimes are so long and testing and/or operating experience is so limited.

Reliability is enhanced by the design of redundant systems and the ability to perform maintenance in orbit. Table I shows, for typical subsystems, the effects on reliability and weight of these alternative approaches in a 1-year space station.

Thus, in this instance, for an investment of 444 lb of spare parts, overall reliability can increase from 0.2% to 97%. It is interesting to note that the subsystems which show the most urgent need for spares and maintenance are the environmental control and life support subsystem and the stability and control subsystem, which are discussed later in this paper.

An evaluation of the effect of maintenance and redundancy can be conducted in another domain. In Figure 2, the weight of three alternate approaches to the design of a single subsystem is shown as a function of mission time. In Curve A, a single repairable subsystem is assumed; the weight increase with time is a result of the introduction of spares as failures are experienced. Curve B represents a repairable redundant design approach for the same subsystem. The initial weight of this subsystem and weight increase per unit time is greater than the single system because two sets of components are involved in the maintenance procedure. Curve C represents a redundant system designed with a high-enough intrinsic reliability to preclude the necessity for maintenance. For the subsystems illustrated, the appropriate design might well be B because C involves undue weight penalties and A, unacceptable down-time. If the repair implied in the selection of the case shown by Curve B involves crew skills not readily available or replacements or checkout not easily accomplished in orbit, then the heavier design, Curve C, must be used.

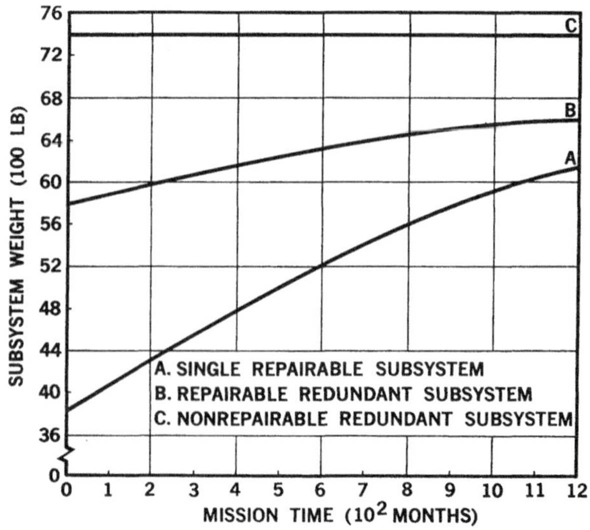

Fig. 2. Reliability and weight optimization. Electric power design options. Solar cells, 6 kW nominal.

New techniques to enhance reliability are being considered by subsystem and system manufacturers in the United States. The question is being asked by Radio Corporation of America (RCA), for example, as to whether functionally standardized subassemblies can be employed in diverse subsystems to ease the logistics and spares storage problem. Can a standard analog-to-digital converter card, for example, be used for both the orbital computer and data handling system? If this multiple application can be achieved, then replacements parts might be drawn, on an emergency basis, from lower priority subsystems. Interconnection of some functional elements in diverse subsystems might give the station a great versatility and high reliability.

3. Emergency Crew Return

In the most general terms, four responses can be made to emergency situations in orbit. These are as follows:

(1) On-board repair of the malfunction.

(2) Activation of on-board emergency survival systems.

(3) Initiation of space station escape and return systems.

(4) Ground-initiated retrieval.

These responses are interrelated. For example, the nature of a ground-initiated retrieval system depends heavily on the amount of inherent emergency survival time in the spacecraft.

In the United States, various agencies have studied these responses to orbital emergencies. The ability to repair malfunctions in orbit implies that the space station carries tools and spare parts. Optimal selection of tools and spare parts depends, in turn, on systems analysis of potential malfunctions. Obviously, situations in which repair is an acceptable emergency response are limited to the following:

(1) Those in which the malfunction has not destroyed essential subsystems of the space station.

(2) Those in which the malfunction has not obviated the astronaut's ability to manipulate controls and tools.

(3) Those in which the malfunction has been anticipated in stocking on-board spare parts.

(4) Those in which items which have failed are accessible and repairable.

Emergency survival requires the use of a compartmentalized space station. For example, if a high-energy meteoroid were to penetrate a compartment and cause rapid decompression, astronauts in the surviving section could close the hatch as is done when a submarine compartment floods.

Ground-initiated rescue describes the situation in which a potentially disabling malfunction is discovered in the orbiting spacecraft and a rescue vehicle must be launched from the ground. The orbital mechanics of ground-initiated rescue are extremely complex, but an analysis of the rendezvous problems for any orbital inclination can lead us to a rather simplified set of cases for low earth orbit (Figure 3). If a spacecraft in a 200-nmi orbit at various orbital inclinations is considered and the performance of the available boosters in a ground-initiated rescue mode is evaluated, it appears that orbital survival time can be as low as 5 hours if the Saturn V vehicle is used, or as high as 17 hours if the smaller vehicles are used. This figure depicts the relationship between booster capability and orbital survival time at two orbital inclinations: 28° and 60°. In computation of the data for this figure, no range-safety restriction to the launch was assumed, and the vehicles were launched within 1 hour of the emergency. The only approach which meets the short-response-time criterion is the on-board space station escape system.

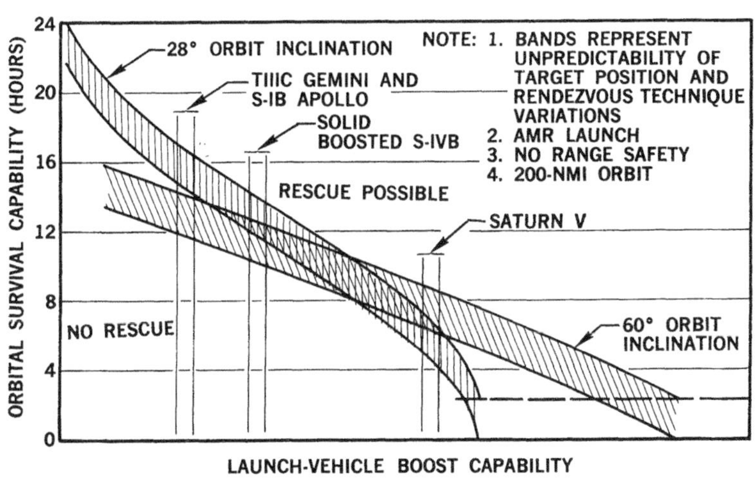

Fig. 3. Earth orbital rescue.

The escape response to an emergency situation implies that the space station has an orbital analog of the parachute on-board. The most easily envisioned escape system involves the use of re-entry modules such as Gemini and Apollo which could have been boosted to orbit as a portion of the original space station or as a component of a subsequent cargo resupply vehicle. These modules could remain attached to the spacecraft in a 'ready alert' state.

4. Man Performance at Zero *g*'s

Of the problems described in this paper, least is known about man, his reliability, functioning, and interface with other systems in the orbital environment. Man's role

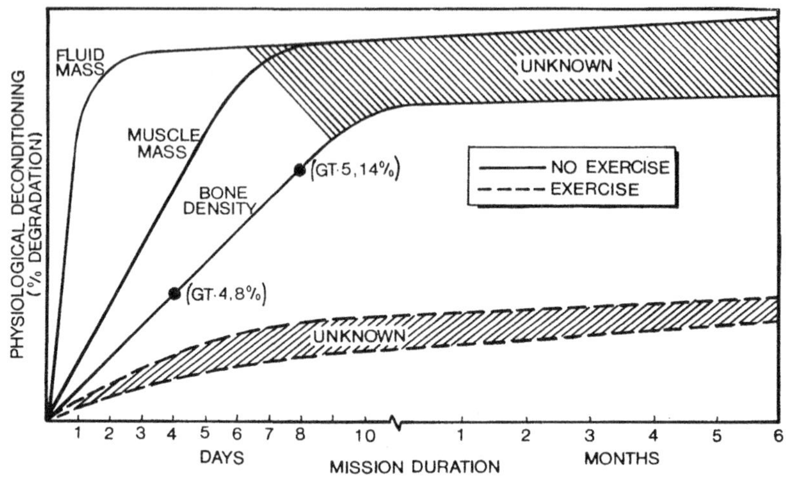

Fig. 4. Physiological effects of zero *g*.

in the mission will require that he operate and maintain the vehicle and equipment; furthermore, he will serve as both an experimenter and as a subject throughout the mission. So that man can function properly in this dual role, it will be necessary to match personnel and equipment to make appropriate use of human capability without undue physiological or psychological compromise. Maladaptation of critical body functions to prolonged zero-*g* exposure has been considered in past space station studies. The cardiovascular and musculoskeletal systems are potentially involved in reactions to zero-*g*. Figure 4 illustrates deterioration of some basic body functions, based on the extremely short exposure data provided by the Gemini Program. As shown, fluid mass maladaptation onset appears faster than muscle mass deterioration or decalcification. Adaptive changes of these human systems will occur and will necessitate establishment of new standards of performance consistent with the environment. Although early manned flights have not indicated whether physiological deterioration at zero-*g* will be a problem, present opinion generally holds that therapeutic and preventative regimens will satisfactorily mitigate potential difficulties. One of the major purposes of first-generation space stations will be to investigate this problem in greater depth and over longer periods of time.

With these regimens in effect, the problem will be not whether man can exist in space but how he will perform over long periods. This problem becomes critical when considering space station design because the remedial techniques which must be integrated into the habitability features provided for the flight crew can dictate widely variant approaches to station design.

McDonnell-Douglas studies have indicated that intermittent exposure to centrifugation should assist in preventing some aspects of cardiovascular deconditioning. Specifically, the centrifuge applies a head-to-foot gravity gradient to stress the cardiovascular system. The centrifuge will also be invaluable in conditioning crew members for re-entry, and in simulating re-entry gravity profiles. It is anticipated that a routine centrifuge schedule can be developed which will require no more than a half-hour per man per day. A variety of active and passive exercise techniques will also abate physiological deterioration. These techniques counteract not only the effects of weightlessness, but those resulting from lack of normal muscular work load as well.

Rotation of the whole station will also produce an artificial-gravity environment but may disturb a large portion of the mission experimental program. Inclusion of counterrotating or detached zero-*g* modules can obviate this problem but complicate station design. However, it may prove more desirable to rotate the entire station than to experience loss of crew effectiveness in long-term zero-*g* operation. The physiological limits of space station rotation are depicted in Figure 5. At the upper bound, of

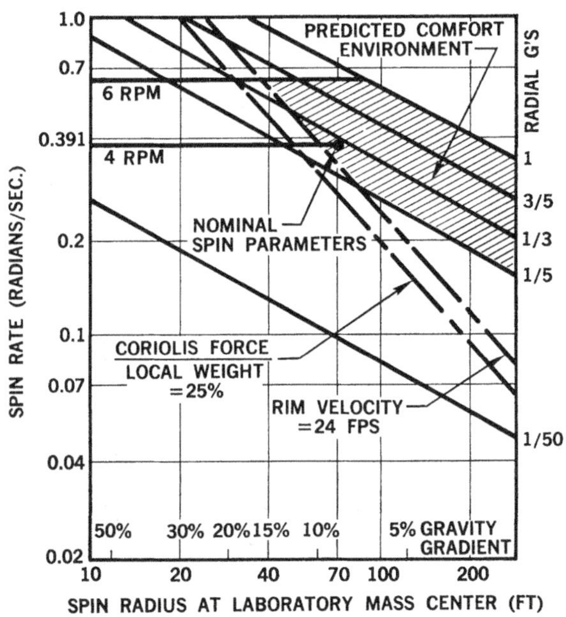

Fig. 5. Artificial-gravity parameters.

course, is 1 g, and rotational speeds which may bring about canal sickness from Coriolis effects. The lower bound is usually taken to be about $\frac{1}{5}$ g, which is required to provide adequate friction for locomotion. Radius arms of greater than 40 ft are generally thought to be required to produce gravity gradients below 15%. The final bound depicted on this curve is based on space station rim velocity; at too low a velocity, an astronaut's walking counter to rotation would reduce the gravitational levels appreciably below those intended. This figure shows that radius arms of about 70 ft are generally desirable; however, this distance may be difficult to achieve in practice with configurations being contemplated today.

Determination of the deterioration of crew effectiveness at zero-g and the demonstration of therapeutic and preventative techniques awaits testing in an operational orbital environment.

5. Stability and Control System

Stability and control system (SCS) requirements are formed primarily by the needs of the experiments, but must include consideration of the spacecraft orientation, and static and dynamic disturbances experienced by the space station.

Figure 6 depicts several of the orientations possible in a long-duration mission. The belly-down orientation shows to advantage for long-term orientation. In this orientation, the long axis of the vehicle is aligned with the velocity vector; this requires that the vehicle complete one revolution about its axis for each revolution about the earth. Earth orientation sensors will continually point at the earth, without gimballing; gravity-gradient torques are also minimized by this orientation with resultant propellant economy. The inertial orientation is required by astronomical experiments.

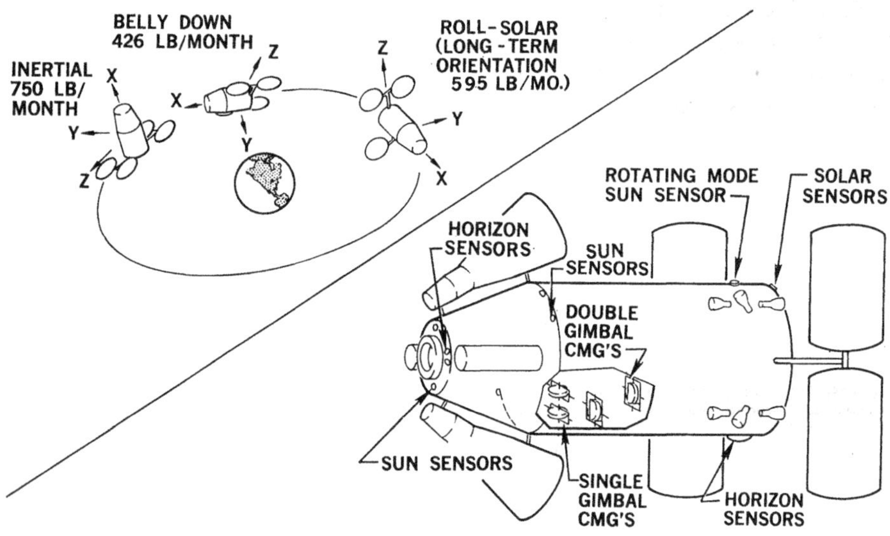

Fig. 6. Stabilization and control system.

In this orientation, the long axis of the station is kept fixed in space as it rotates around the earth. In this case, cyclical gravity gradient torques are generated which must be sensed and compensated by the control system.

Some of the major SCS components and their locations are shown at the lower right of Figure 6. Horizon and solar sensors, in combination with an inertial package, can provide the primary attitude information. Integrating rate gyros are used to provide short-term stabilization. Control moment gyros (CMG's) can be used as primary actuators by storing angular momentum and contributing this momentum to

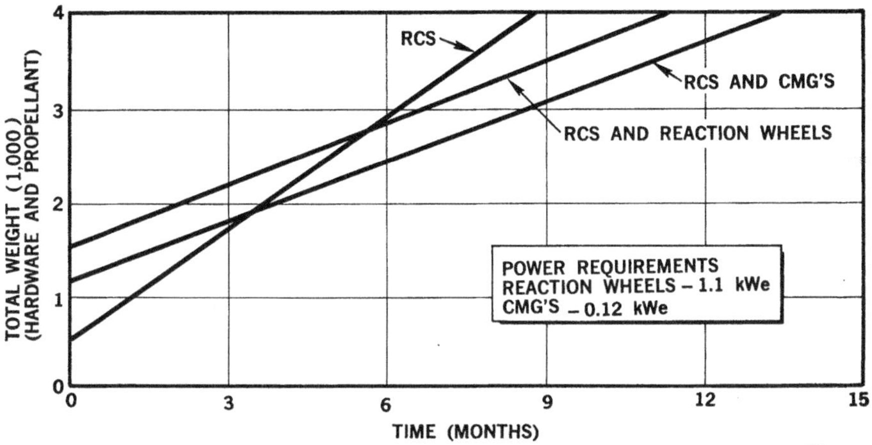

Fig. 7. SCS comparison.

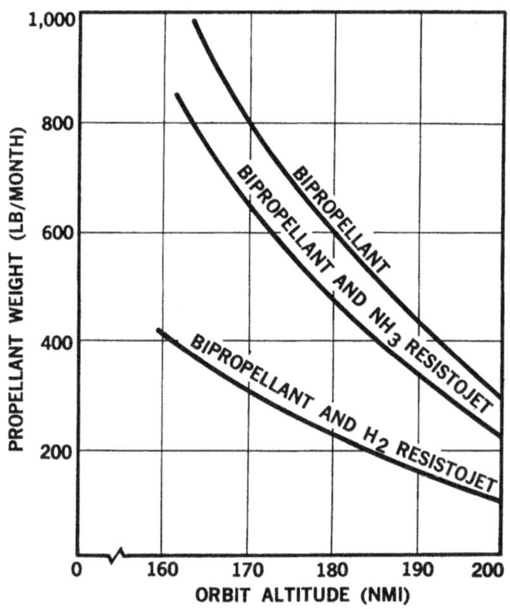

Fig. 8. Attitude-control and orbit-keeping propellant requirements.

the station to cancel motion resulting from disturbing torques. This system conserves attitude control propellants which might otherwise be required; with CMG's only small, long-life reaction control engines need be used because their primary function is to unload the gyros near saturation. Figure 7 shows the weight advantage which accrues to the use of control moment gyros with a small reaction control system for six-man space station; after approximately 3 months, the CMG system requires less total weight than an attitude control system based on reaction thrust alone.

The presence of aerodynamic drag requires that force be added to the space station so that orbital altitude is maintained. This force can be provided by periodic firing of high-thrust engines, fueled with bipropellant systems or with high-efficiency low-thrust systems, such as the resistojet. Figure 8 shows the total system weight incurred by these alternative approaches as a function of the space station's orbit altitude. Although hydrogen resistojet system functions with the highest specific impulse (720 sec), ammonia reactant appears more desirable when power requirements, tank diameters, and storage requirements are considered.

Finally, it is possible to determine optimum orbital altitude from the standpoint of reaction control system performance. Figure 8 shows the weight of propellant required for orbit keeping and attitude control as a function of altitude and specific impulse of various types of reaction control systems. Figure 9 shows the loss of payload weight experienced by a typical logistics vehicle as altitude is increased. From these two considerations, a family of optima may be derived in which the discretionary payload available in the logistics craft is maximized for the particular reaction control system. This family of optima is shown in Figure 10.

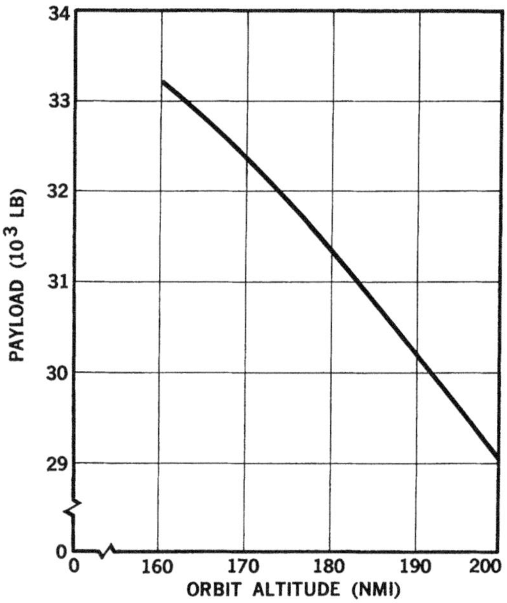

Fig. 9. S-IB payload capability to circular orbit (50° inclination).

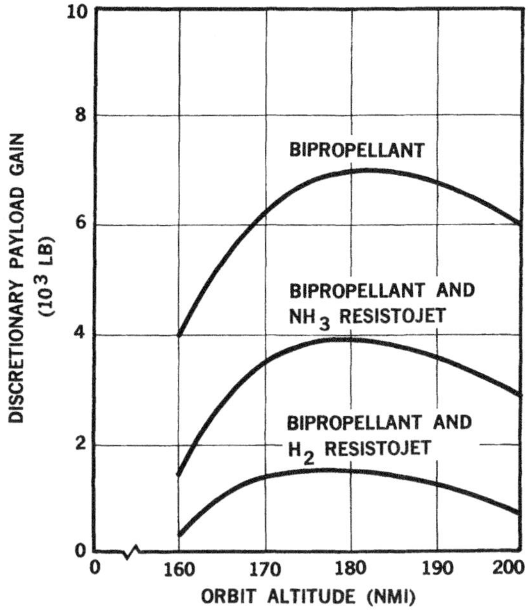

Fig. 10. Orbit altitude optimization.

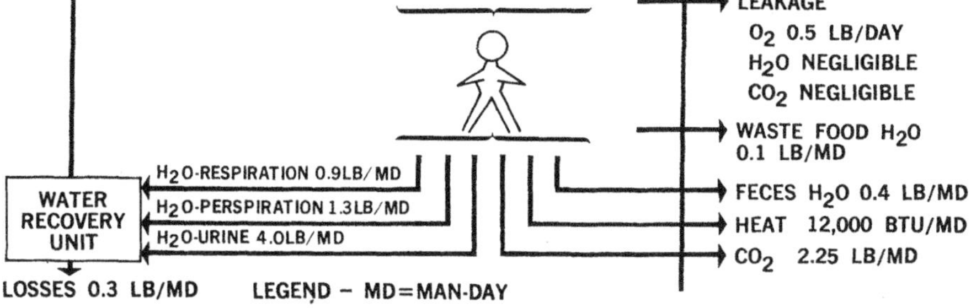

Fig. 11. Metabolic parameters.

6. Environmental Control and Life Support Systems

Environmental control and life support (EC/LS) system requirements are dictated by the crew size, station operating time, and crew rest/work cycle. A typical work/rest cycle for a 9-man, long-duration station results in metabolic balances depicted in Figure 11. Man's requirements are food, water, and oxygen, and his output is solids, expired gases, and liquids. As indicated, approximately 9.4 lb of material are processed by each man each day. This material can be provided in several ways, as follows: 'open-loop', that is, stored, used, and expended; 'closed-loop', that is, stored, used, and reprocessed to reuse; or in combinations of these approaches. Open-loop systems are simpler but less efficient. The evolution toward closed-loop systems is shown in Table II.

TABLE II

Evolution of environmental and life support systems

System	Gemini	Apollo	First-generation space station	Long-duration space station
Atmospheric supply	5 psi supercritical O_2, no diluent	5 psi supercritical O_2, no diluent	7 psi 50–50 N_2 O_2 (subcritical)	7–15 psi with 20% O_2 (closed cycle)
Atmospheric purification	LiOH charcoal filter	LiOH charcoal filter	Regenerable molecular sieve, contaminant control	Semiregenerable molecular sieve, contaminant control
Water management	Drink fuel-cell water	Fuel cell water	Partial closed H_2O (humidity)	Closed H_2O (humidity and urine)
Waste management	In space suit store unprocessed	In suit toilet store unprocessed	Toilet and processing for storage	Toilet and processing for storage
Compartment conditioning	Cabin circuit	Cabin circuit	Cabin circuit	Cabin circuit
Atmospheric pumpdown	N/A	None	None	Yes
Food	Semi-liquid, paste, etc.	Semi-liquid, paste, etc.	Wet food and freeze-dry	Wet food and freeze-dry

Consideration of the possibility of inadvertent atmospheric contamination, depressurization, or fire make it desirable to include a minimum of two isolated compartments within the pressurized area of long-duration space stations. Figure 12 represents a typical integration of an EC/LS into a space station. Because of the two-compartment philosophy, duplication of the atmosphere purification, suit-conditioning subsystems, and the ventilation and temperature controls is required. Atmospheric purification must be duplicated in the experimental laboratory because of biological and/or fluids experimentation.

From a physio/psychological standpoint, duplication of the earth's atmosphere is desirable; however, leakage from the station can be expected to increase at higher atmospheric pressures. Furthermore, at higher atmospheric pressure the possibility of dysbarism is introduced when the crew moves from the space station environment to a lower-pressure oxygen-rich chamber. Current space suits and logistics craft operate at 3.7 psia oxygen.

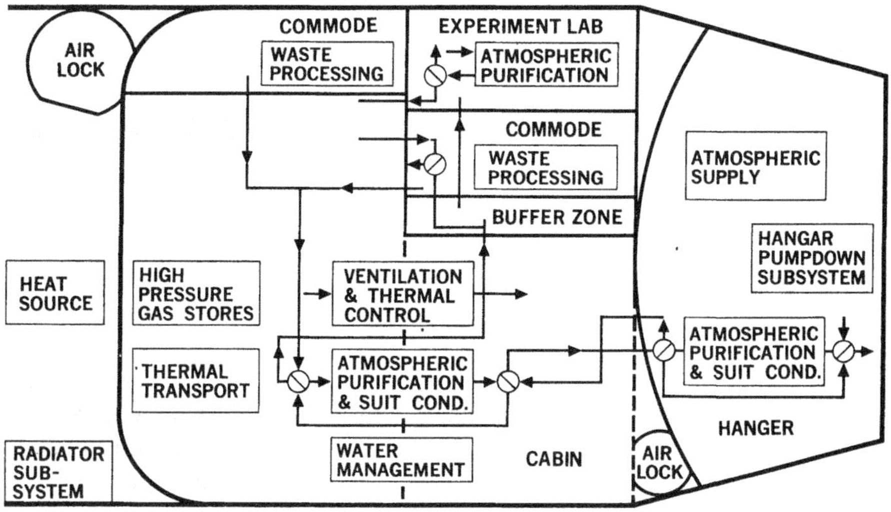

Fig. 12. EC/LS subsystems.

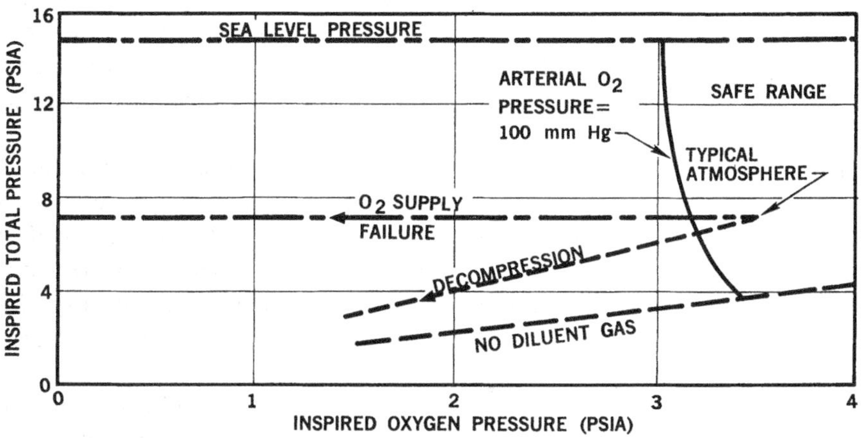

Fig. 13. Physiological limits for O_2 arterial O_2 and functional impairment.

If space station pressure is to be less than 14.7 psia, the overriding consideration in minimizing physiological deviations is to provide an atmosphere which will result in a blood saturation of oxygen approximately the same as that at sea level (approximately 100 mm Hg or 1.94 psia). The required, inspired, oxygen partial pressure versus total inspired pressure for various levels of the arterial oxygen tension are shown in the curve in Figure 13. The curves show that the inspired oxygen partial pressure required to provide an arterial tension of 1.94 psia varies from 3 psia at sea level total pressure, to 3.46 psia at 3.7 psia total pressure. With the oxygen partial pressure thus bracketed, a dilutent and a total pressure must be selected; nitrogen or other inert gases can be used as a dilutent but these gases introduce the possibility of dysbarism when the crew is required to move from an area of high pressure to a lower total pressure, as occurs when the crew changes from space station to pressure suit or to a logistic vehicle. The point of inception of dysbarism varies among individuals and for individuals at different times. The N_2/O_2 pressure ratio below which bends are not a problem is 2.0. To minimize fire hazard, the ratio of nitrogen to oxygen should be as high as possible. Selection of the total cabin pressure is then bounded by a compromise of dysbarism and fire hazard on the upper bound and the cardiovascular oxygen saturation on the lower bound. Distribution of atmospheric pressure also affects the pressurized compartment weight, fan power, and weight leakage losses. Consideration of all of these factors generally leads to a choice of 7 to 10 psia.

The atmospheric supply, purification, and water-management systems are all interrelated, in that perturbation of one affects the design of all. As shown in Table II, the Gemini and Apollo systems used supercritical oxygen storage systems. However, when six- to nine-man space stations are considered, O_2 storage requires development of large-size, thermally insulated tanks, and perhaps, oxygen resupply. The advantage of a closed oxygen system is shown in Figure 14 for a typical 6-man space station operating for 5 years. Present methods of oxygen regeneration include the Bosch, Sabatier, and molten carbonate processes. However, all require complete or partial closure of the water-management system to be effective. Figure 15 typifies a Bosch oxygen regeneration system wherein oxygen is provided by electrolizing water to produce breathing oxygen. The hydrogen produced is directed to an accumulator in the

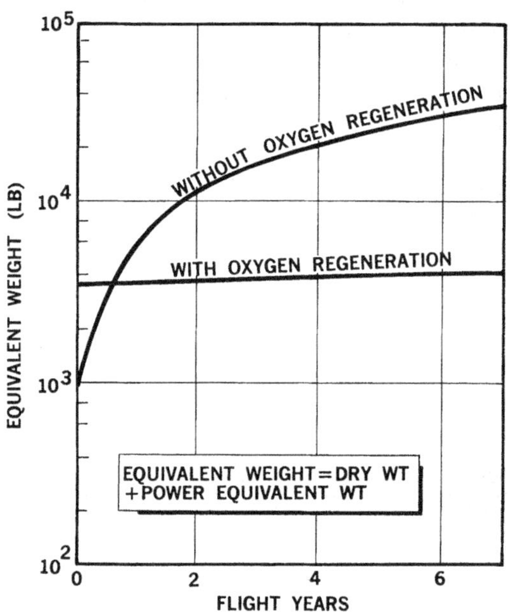

Fig. 14. Oxygen.

carbon dioxide reduction subsystem. The water used to provide consumable oxygen can be supplied by a logistics vehicle on a regular schedule. Further sophistication of the water-management system would preclude use of this resupplied water.

The Bosch process for CO_2 hydrogenation, as given by

$$CO_2 + 2\,H_2 \rightarrow 2\,H_2O + C$$

is desirable because it is a one-step process. An alternate oxygen regeneration process is the Sabatier with methane conversion to acetylene, as shown below:

$$CO_2 + 4\,H_2 \rightarrow CH_4 + 2\,H_2O$$
$$2\,CH_4 \rightarrow C_2H_2 + 3\,H_2$$

The hydrogen required by these reactions can be replaced if a high-efficiency water-

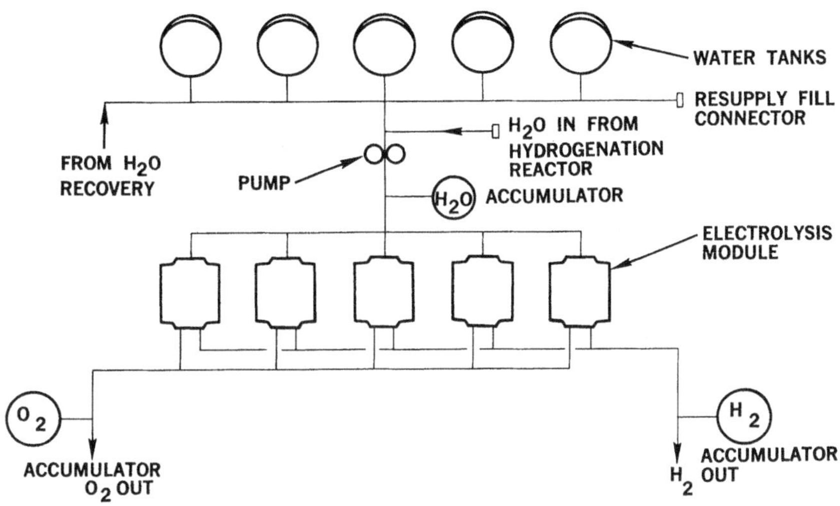

Fig. 15. Atmosphere supply subsystem.

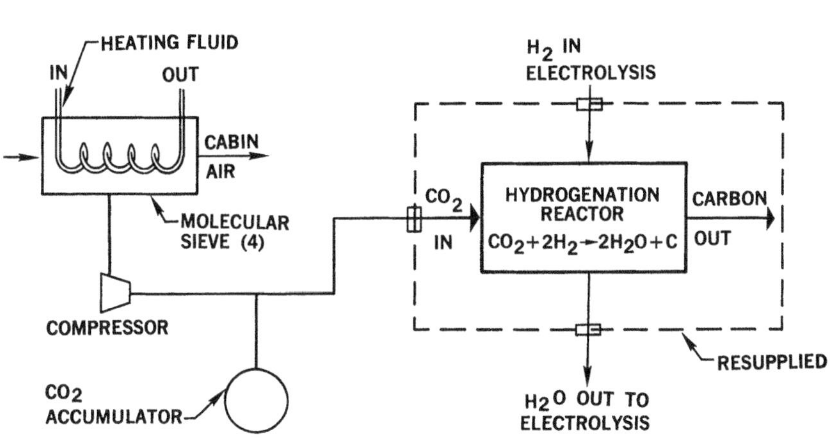

Fig. 16. CO_2 collection and reduction.

recovery process, such as air evaporation, is used. Hence, the key to the Sabatier system is essentially a 100% water recovery from urine and humidity water. The thermal load requirements for this oxygen regeneration are approximately 6 kW for a 9-man system as compared to approximately 3.4 kW with an open oxygen cycle system.

Although CO_2 removal from operational spacecraft has been adequately handled by LiOH, its use in large space stations for long durations exacts a large weight/volume/maintenance penalty; hence, semiregenerable or fully regenerable molecular sieves are required for long-duration space-station operation. Figure 16 shows a typical CO_2 collection system using molecular sieves. In any of the reprocessing systems, wastes might be used as propellants for advanced low-thrust reaction-control systems, for example, resistojets.

Water management has not been a problem on the Gemini and Apollo programs because the water generated by the fuel cell was adaptable. However, as will be later discussed, long-duration space stations will probably use solar cell or isotope power supplies; hence, water must be supplied or generated. The first approach to closing the water cycle would be to recover the respiratory and perspiratory water by use of multiple filters. The next, more complex, step would be the recovery of wash water. Recovery of water from the molecular sieves used in the CO_2 collection system is possible, but this requires development of fully regenerable molecular sieves. Complete water system reprocessing involves urine and fecal water recovery which requires either a vapor-compression unit or an air-evaporation system and a catalytic burner for the fecal water. Figure 17 shows the dramatic effect of water cycle closure and CO_2 removal using molecular sieves.

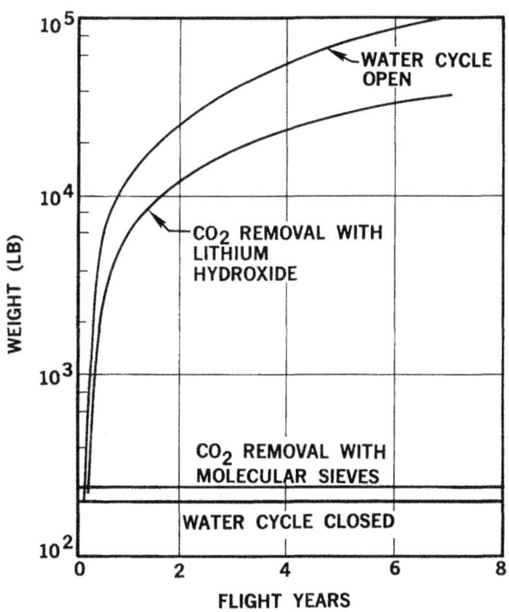

Fig. 17. Water and carbon dioxide.

The most advanced space stations will undoubtedly have EC/LS systems which reprocess all water, and convert the CO_2 produced in human metabolism to usable oxygen. The by-products of this reaction will find use on the station, for example. as attitude control system reactants. The complete closing of the ecological cycle will involve biological processing of solid wastes and photosynthetic conversion of CO_2 and water to oxygen and carbohydrates. The operational use of these techniques lies several decades in the future.

7. Power Subsystem

Power system approaches which have been studied or used include fuel cells, solar cells, static and dynamic conversion systems powered by radioisotope sources, and

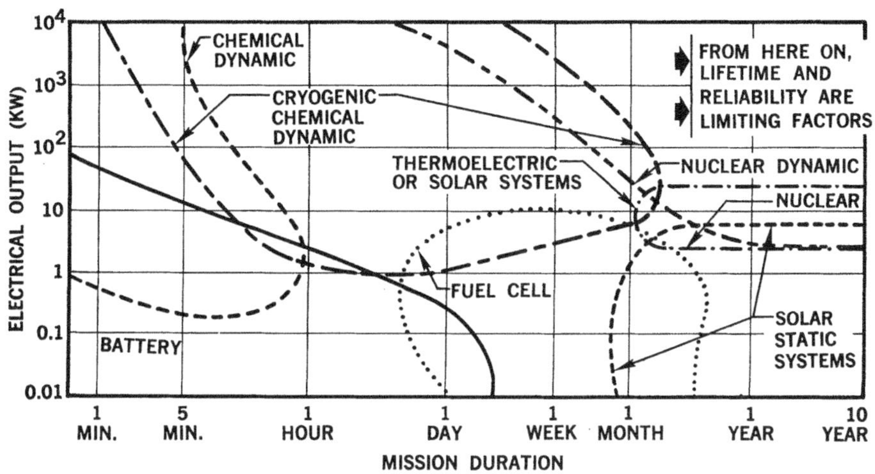

Fig. 18. Potential space power systems applications.

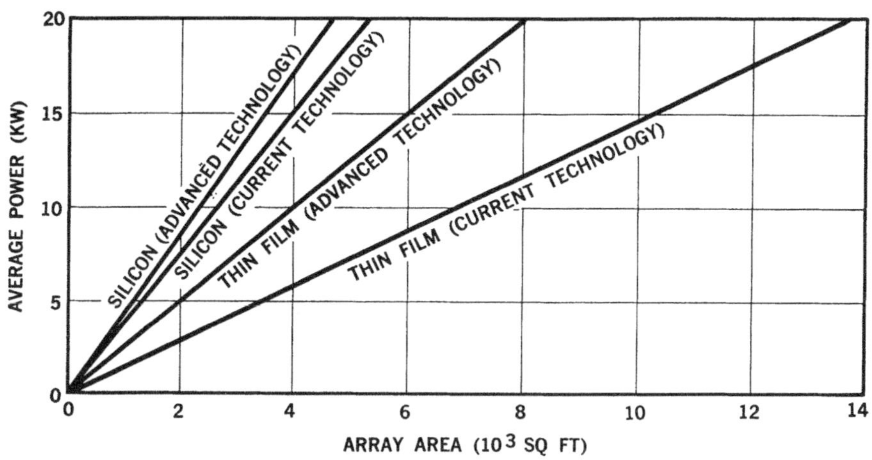

Fig. 19. Solar-cell array area (low earth orbit).

reactor power sources. The choice between these systems depends on the power demands, the mission requirements for the space station, the space station lifetime, and the launch date. Generally, past space station studies have indicated that approximately 2 kW per man should be generated; the environmental control and life support system and experiments are the major contributors to this demand. Thus, the power levels of interest for space stations under study range between 4 and 50 kW, with operating times ranging between 1 and 5 years. Weight and reliability considerations lead to a map in the power level time domain of the applicability of these subsystems, shown in Figure 18. Batteries are practical at only low power levels for short durations; fuel cells extend the regime in both dimensions; solar cells and radioisotope dynamic systems are appropriate at moderate power levels for long durations; and reactors, finally, are the most promising approach to high power levels for long periods.

Solar cells offer great promise as space station power sources, particularly if current work in thin-cell design is successful in producing low-cost arrays of large physical size. The tradeoff between solar-cell array size and power availability for contemporary and forecasted cells is shown in Figure 19. This figure shows that for reasonable power levels, relatively large physical arrays are required. These arrays may be cumbersome to store during launch and erect in orbit; however, current flexible cell developments point to the possibility of 'roll out' development in future stations. These large area panels generate body drag in orbit which must be compensated by the station's orbit-keeping propulsion. The effect of solar panel drag on orbit-keeping propellant weight is illustrated in Figure 20. In addition, solar cells impose re-

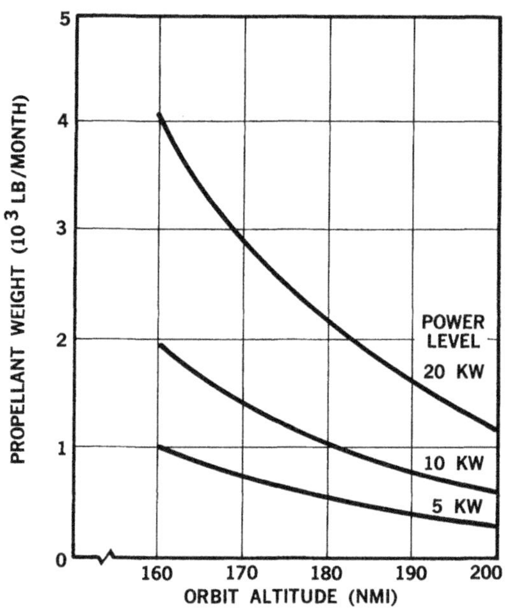

Fig. 20. Orbit-keeping propellant for solar panels.

strictions on experiment viewing directions and require auxiliary power of some sort for the period of operation in the earth's shadow. It has been found that rechargeable batteries generally satisfy this requirement. The attitude control reaction jet exhaust produced must be physically separated from the solar cell panels to avoid deposition and concomitant loss of efficiency.

Fuel cells require cryogenic storage of reactants and require logistics support if operating time in orbit is to exceed several months. The effect on logistics of fuel cell

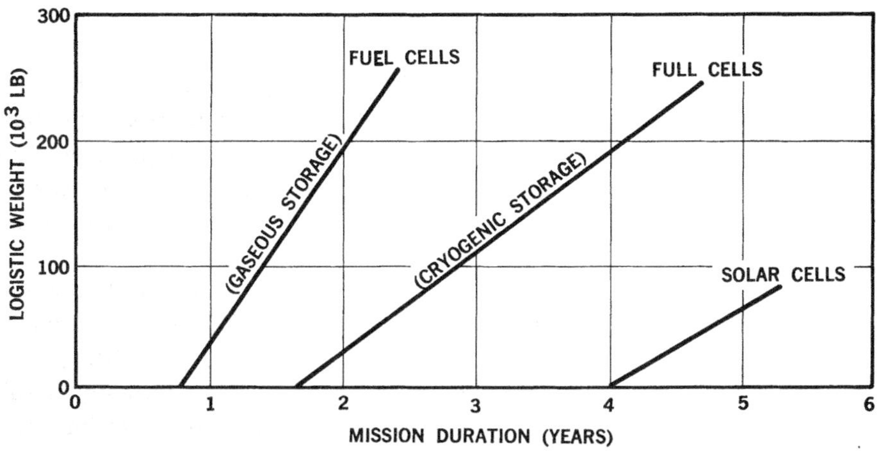

Fig. 21. Logistics-fuel cells vs. solar cells.

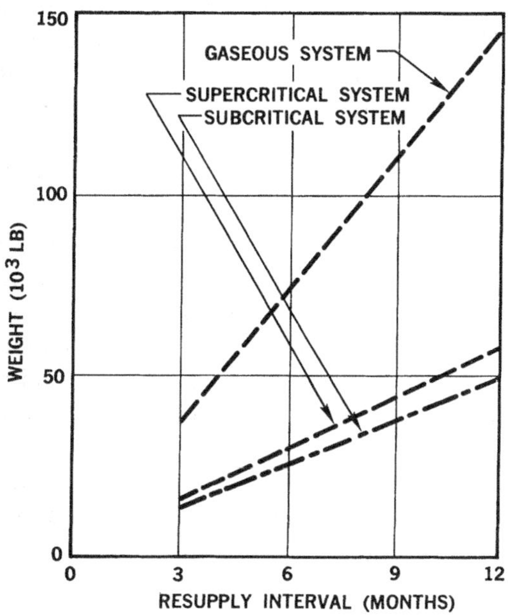

Fig. 22. Reactant storage weight vs. time.

power compared with solar cells is shown in Figure 21. The weight required for various modes of storage of the fuel cell reactants is shown in Figure 22. Of the methods currently available, subcritical storage appears to be the lightest system; however, storage time of available bottles is limited to approximately 90 days. Methods have been proposed to extend storage times to 1 year and this appears to be the practical limit at present. Although fuel cells are complex, they offer the advantage of potable water production as a by-product of the power-generation process. For example, three advanced Apollo fuel cells generating a total of 2700 W of power produce 59.6 lb of water per day. If other power systems are selected, potable water must be provided by other means.

The plutonium 238 (PU-238) fueled isotope Brayton cycle system has been recommended for long-term space stations. The installation contemplated is shown in Figure 23. However, radio-isotope Brayton systems are expensive and the rotating machinery portions of the system have unproven reliability. Of the isotope systems which have been considered, the PU-238 Brayton cycle is generally regarded to have the highest efficiency (Figure 24).

Reactor power systems offer the potential of satisfying large power requirements but present a significant shielding problem with regard to the crew. If dynamic conversion systems are used with the reactor, the space station crew will probably find it necessary to replace the machinery every $2\frac{1}{2}$ years or so. This operation will be delicate. However, static systems that use thermoelectric conversion may eliminate the need for this orbital replacement of the power-conversion system. Figure 25 shows the relationship between power level and subsystem weight at a given dosage level for various distances between the space station and the power system.

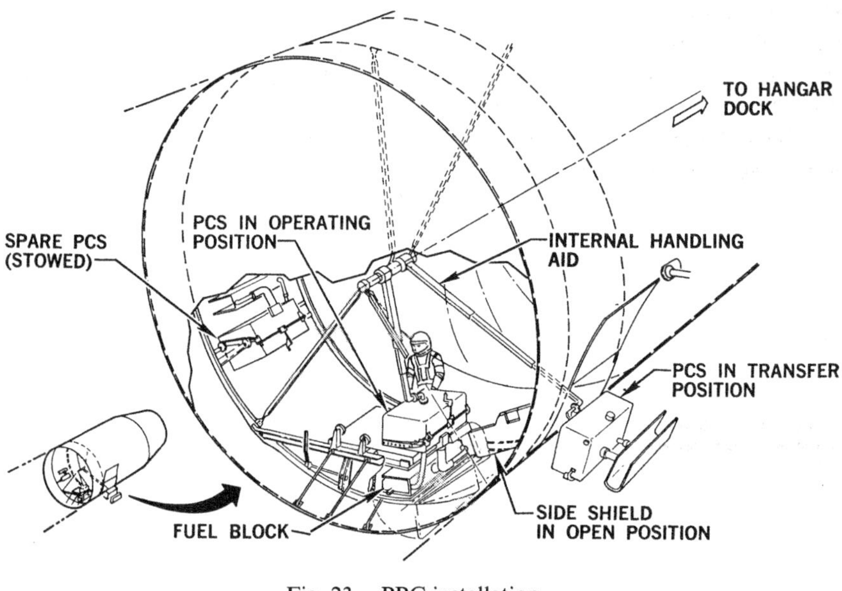

Fig. 23. PBC installation.

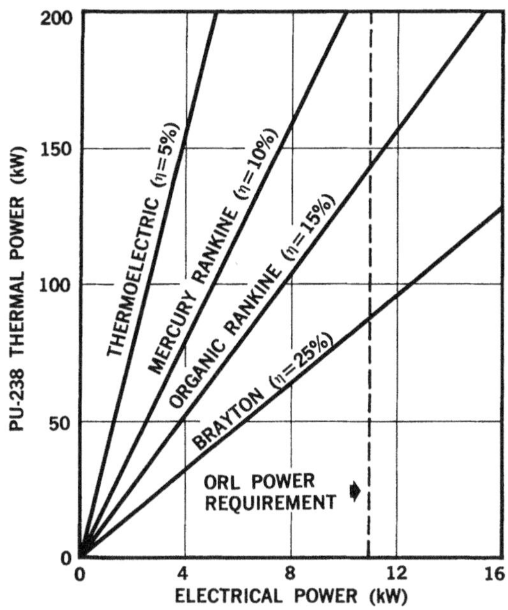

Fig. 24. Radio-isotope systems efficiencies. Note: fuel for two space stations.

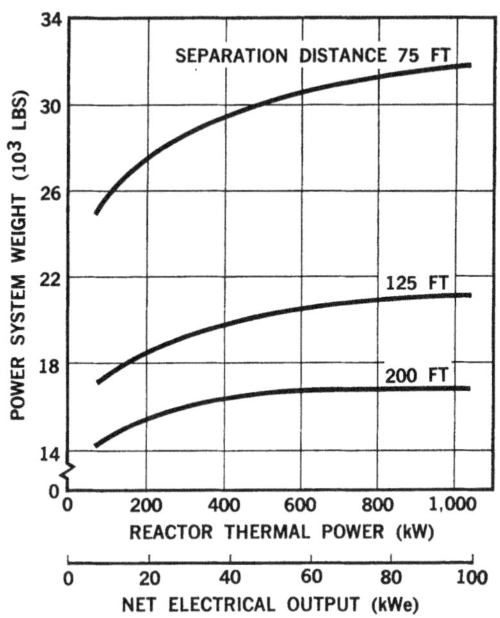

Fig. 25. Reactor power system weight vs. power level.

TABLE III

Power source comparison
(1-year integral launch)

Power source	Launch weight (lb)	Reliability	Development risk	Remarks
Fuel cells/battery:				Limited growth potential, high launch weights, potable water for crew
(A) supercritical storage	72 500	Fair	Medium	
(B) gaseous storage	157 000	Good	Very low	
Solar cell/battery	4500	Average	Low/medium	Potential problem with gimbals and deployment, operational/experiment interference
Solar cell/fuel cell	116 000	Good	Low	Complex, limited growth, operational/experiment interference
PU-238 isotope Brayton cycle	4000	Average	Medium	Nuclear safety, isotope recovery and/or disposal

Table III contains a comparison of fuel cells, solar cells, and isotope systems for a typical 1-year space station without logistics, and with consideration of the reliability and availability of these approaches.

8. Conclusions

Although several potential engineering problems of the long-duration space stations have been discussed, the *absence* of a large number of major problems is really significant. The three critical problems can be solved: long-term reliability, crew escape, and man's performance at zero g's. Some directions and alternatives are already becoming clear through our space station studies (see article on 'Manned Space Stations', by R. R. Gilruth, p. 1). The answers these studies provide will shape the form and influence the schedule of the long-duration station.

THE RESEARCH POTENTIAL OF MANNED EARTH ORBITING SPACECRAFT FOR METEOROLOGY

S. FRED SINGER

Department of the Interior, Washington, D.C., U.S.A.

Abstract. Manned satellites can advance atmospheric science in a number of important ways. (1) Develop more rapidly and economically a refined system of electromagnetic sensing instrumentation which can be employed on future operational weather satellites. (2) Carry out specialized observing programs; e.g., in connection with earth-based storm modification experiments. (3) Perform a maintenance and repair function for the advanced weather satellites in a future operational system.

A human experimenter can make unique contributions in devising instrumentation to measure night-time cloud distribution, winds, temperatures, precipitation, water vapor and ozone content, cloud properties, aerosols, and the state of the sea, of snow, ice and terrain.

The economic value of an advanced weather satellite system in terms of improved weather forecasts is pointed out; it may also help solve scientific questions about the changes in climates.

1. Atmospheric Data: Why, What Kind, and How Much

The earth's atmosphere is a complicated physical system. We believe that we understand many of the physical laws which control what takes place in the atmosphere: the motions of air masses, the evaporation and condensation of water, even some of the microscopic processes in the formation of cloud droplets, and the less well-known problems of atmospheric electricity. Why then can we not predict the state of the atmosphere as easily as we can predict the occurrence of astronomical events?

There is an important difference between a system such as the atmosphere and a system such as the planetary system. The latter is relatively simple. The motions of the planets are controlled by gravitational forces only, and because of the great distance between the planets, the interactions are basically between the sun and a single planet with any other effects entering in as minor perturbations. In the case of the planetary system, therefore, it is possible to perform one set of observations at a given time and then extrapolate or *predict* the motions of planets from then on. In principle the same procedure of numerical prediction is possible in the atmosphere, but in practice it is not.

The complication of atmospheric interactions is such that frequent observations are essential, and the predictions that can be made are good only for a very short time. Our aim, of course, is to increase the length of the prediction, and this problem of being able to predict the atmosphere is taken to be synonymous with our 'understanding of the atmosphere'. If we 'understood' the atmosphere perfectly, then we should be able to take one set of observations at a particular time and from then on be able to predict the course of atmospheric processes, including all of the weather phenomena all over the globe.

More than any other science, meteorology relies on observations. Both the quality and the scale of these observations are important. By 'quality' we mean the physical

S. F. Singer (ed.), Manned Laboratories in Space.

quantity which is being measured; i.e., the atmospheric pressure or the relative humidity, and the precision to which it is measured. By 'scale' we understand the fineness with which these observations are carried out either on a horizontal scale, or on a vertical scale, or on a time scale.

A satellite can be considered as a meteorological observing tool *par excellence*, which surveys the whole globe daily. This time scale is adequate for most meteorological tasks; the horizontal scale can, in principle, be made as fine as necessary.

There is another parameter that one is usually interested in and that is cost. Meteorological observing systems are expensive to set up and expensive to maintain. Eventually when we have to make a decision about whether a satellite is a *better* way to observe the atmosphere, we must also consider the cost factor.

It has been adequately demonstrated that a satellite represents the most economical method for getting regular global data of a certain kind, e.g., the day-time cloud distribution. It is less well recognized that once improved instrumentation is available, much more data can be obtained at relatively little extra cost. An example, now within technical range, is an increase in horizontal resolution. Much more extensive development is required, however, before different types of observations can be perfected: the cloud distribution at night; the altitude and thickness of the cloud; its detailed properties; precipitation, etc.

2. The Advantages of a Manned Platform

It is the major contention of this essay that a manned space vehicle provides an economical method for investigating the limits to which satellite meteorology can be pushed. It is not intended that the manned orbital laboratory itself be an operational system for observing the weather routinely, but that it be used to advance the state of instrumentation of weather satellites by developing an elaborate system of sensors which could be used in unmanned meteorological observational platforms of the future, as the second or third generation of today's meteorological satellite system.

It should be noted that the man is not used as the operational observer; that is left to instruments. His function is three-fold: (1) to act as the experimenter in developing efficiently some very advanced instrumentation; (2) to be used for specialized non-routine types of observations (e.g., during storm modification experiments); (3) to maintain or repair instruments in an elaborate meteorological observatory of the future.

The use of a man, rather than a robot, in all of these operations must be established on the basis of cost. It can be shown quite directly that as complexity increases reliability of operation must be purchased at an ever increasing cost. Beyond a certain level of complexity the use of a man becomes almost mandatory, not only in space, but in an earth-bound operation also.

Scientifically, the manned platform is desirable because the basic problem of atmospheric observations is such that one cannot predict in advance the exact details of instrumentation which will allow the most beneficial types of observations; i.e., even today it is quite difficult to decide such broad questions as the use of infrared versus microwave sensors for measuring the vertical temperature structure of the atmosphere, and indeed no final decision has as yet been made. Further than that, however, such details as the setting of the exact wavelength, the width of wavelength intervals, and the effects of various atmospheric conditions on the observations cannot be gauged sufficiently well in advance.

Some reflection will show that it is too expensive to develop the necessary instrumentation in unmanned flights, and too time consuming. The reason, of course, is that these unmanned flights would have to be successive so that each one could take advantage of the results of the preceding flight. Such a program extending over a ten-year period would certainly run in excess of 200 million dollars. Yet within this period we will surely have available manned platforms which can carry out the necessary experiments with much less effort, at a much smaller cost, and much sooner.

The economic argument is simple. Orbiting manned platforms will be launched in the near future. It is clear, therefore, that one should not assign all of the cost or even a major fraction towards its meteorological application. It is suggested, therefore, that a proper cost comparison should use an *incremental* cost of the manned program, and in addition assign a value to time gained in the development of instrumentation through the use of experiments on manned platforms.

From experiments on manned platforms an ultimate unmanned weather satellite system could be designed which will allow us to obtain the optimum operational data needed for long-range weather prediction and for the many other tasks which weather satellites are capable of doing.

What then are these various types of measurements? Rather than prepare an exhaustive list, we propose to discuss here those measurements which have important meteorological and other economic applications while at the same time being challenging from a scientific point of view. We will discuss the use of satellites for establishing the cloud distribution by day and by night, for measuring winds at various atmospheric levels, for determining the vertical distribution of temperature in the

TABLE I

Desirable characteristics of manned satellite laboratory

Orbit:	
Altitude	200 n. mi.
Inclination	highest obtainable
Duration	30–60 days
Spacecraft	
Resupply	not needed
Volume for experiments	500 $ft.^3$
Weight for experiments	2000 lbs.
Power for experiments	2 kW
Peak power	5 kW for 1 hr.
Stabilization capability	within 1° of vertical

atmosphere, for measuring the existence and amount of precipitation and water vapor content, for measuring cloud altitudes and cloud properties, for the determination of aerosols and other particulate matter in the atmosphere, for the determination of the overall radiation and energy balance, and for such measurements as sea state, snow cover, and ice distribution. Finally also, we discuss the measurement of atmospheric composition, especially the three-dimensional distribution of ozone.

An effort will be made to introduce into this essay new ideas, some of which have not been discussed previously in the literature. Partly this is because such measurements were thought to be not feasible; however, a manned platform allows us to carry rather elaborate instrumentation, and large weight and power requirements pose no particular practical problem. This feature has a very important consequence which cannot be overemphasized. It allows us to fly and operate a number of experiments simultaneously and use their combined results for drawing conclusions about the atmosphere. The atmosphere is a very complicated system; often to obtain an answer the same question must be asked in a number of different ways.

By contrast, in most unmanned satellites payload space is extremely limited both in terms of weight and power, so that each satellite can only carry one major experiment. As we shall see, the possibility of making measurements with different types of apparatus allows one to obtain a great deal more insight into atmospheric phenomena. And, of course, the presence of a man who is able to make adjustments to the apparatus in the light of the results obtained either through his own intelligence or following the instructions of a ground-based scientific control group can make such an effort much more worthwhile, at a great saving of money and time.

3. Sensing by Electromagnetic Radiation

Before we undertake a detailed examination of the various atmospheric parameters which can be measured by satellites, it is important to examine how a satellite senses the atmosphere. Naturally, a satellite does not make physical contact with the atmosphere; therefore, it cannot use conventional measuring devices such as thermometers, pressure gauges, humidity sensors, etc. Instead it must relay exclusively on electromagnetic radiation which reaches the satellite sensors from the earth's atmosphere below.

All of the electromagnetic spectrum is available but not all portions are equally useful. By and large it makes sense to consider four main *wavelength regions*: the ultraviolet – from about 2500 to 3200 Å; the visible and near infrared from about 5000 to 20 000 Å; all of the infrared from about 2 micron on up to 30 micron; and the microwave region from a few mm to about 30 cm.

A different type of classification is useful, according to the ultimate *source* of the radiation: (1) Scattered or reflected radiation whose ultimate source is the sun or perhaps an artificial source carried aboard the satellite; i.e., a laser or a radar transmitter. (2) Emitted radiation which originates in the atmosphere, or from clouds in the atmosphere, or perhaps from the earth's surface.

TABLE II

Electromagnetic sensing of the earth from satellites

	Ultraviolet (0.2–0.32μ)	Visible and near infrared (0.5–2μ)	Infrared (2–30μ)	Microwave (0.5–30 cm)	VHF ($\sim$ 1 m)
Active					
(Scattered or reflected) (Source may be the sun, or moon; starlight or airglow; or an artificial source such as an optical or IR laser or microwave transmitter)	Ozone distribution	Cloud distribution Cloud top altitude (pressure) Cloud properties Surface properties (terrain, sea state, ice, snow)	Aerosols and particulate matter	Surface properties	
Passive					
(Emitted)	(Airglow)	(Airglow)	Temperature structure Surface temperature Cloud top altitude (temperature)	Temperature structure Water vapor distribution Surface properties	Sferics

Most useful so far has been scattered solar radiation in the visible region which has been used for the TV coverage of cloud patterns and details of the earth's surface. The other important region has been the infrared radiation emitted by the atmosphere, by clouds, and by the earth's surface which has allowed us to determine the temperatures of these three sources.

4. Night-Time Cloud Distribution

A good example of an important experimental task which cannot be laboratory simulated and which is therefore best relegated to an actual experiment in a manned spacecraft, is the design of a television system for night-time operation under very low light levels. It is necessary to use devices which are much more sensitive than the vidicon. For example, image orthicons have been suggested and other more experimental devices. All of these require fairly careful adjustment and tuning, which can be supplied by a man.

But in addition to this, the effects of background radiations such as the airglow cannot be assessed in advance. The problem is one of trying to obtain the greatest contrast between clouds and the earth's surface, and simultaneous elimination of the airglow from starlight, from zodiacal light, and perhaps from a little moonlight.

Different studies have come up with different results; however, a man could be in a position to determine which portions of the electromagnetic spectrum to use in order to eliminate the interfering effects of the airglow and other background emission.

5. Winds

Unfortunately, direct measurement of winds from satellites is difficult, if not impossible. One can conceive of a technique for detecting scattered radiation which is slightly Doppler shifted because of the velocity of the scatterer; but a simple calculation of the magnitude of the shift will convince one that the effect is hardly observable. It is necessary, therefore, to measure winds by tracking something which is carried in the atmosphere. This 'something' can either be a natural particle, perhaps a collection of particles constituting a cloud, or it can be a body artificially placed into the atmosphere, such as a puff of smoke or a floating balloon.

Clouds have the advantage of being available at no cost, but their availability can never be assured. In addition, many clouds do not move with the velocity of the wind; some do not move at all. Great care must be taken before a cloud can be used as a reliable wind indicator. Artificially introduced particles are usually transitory and have to be replaced. If winds have to be measured at several vertical levels and at many locations all over the earth, the number of floating balloons which has to be introduced can become quite large, raising the costs of such a system to a level where its practicality has to be scrutinized very closely.

Much more work needs to be done to recognize the natural particles which can give information about atmospheric motions. It is doubtful whether this can be accomplished simply by taking television pictures of clouds. A manned vehicle has the advantage that the particles can be studied more carefully by examining the 'cloud' at various magnifications by noting their altitude, by studying if necessary the size distribution of the particles, and other important properties.

Even if the cloud method for tracing winds does not become of general operational significance, such studies will enhance our general knowledge about the motion of clouds in the atmosphere and contribute, therefore, in an important way to our understanding of atmospheric processes.

6. Temperature Structure

Far more adaptable to satellite techniques is the determination of the vertical temperature structure of the atmosphere. The method, originally suggested by Lewis Kaplan, is by now well known. It makes use of the fact that CO_2 has strong absorption bands in the infrared. By observing the energy received at a number of wavelengths near 15 micron, one can determine the effective temperature of a number of diffused levels in the atmosphere.

A theoretical study would indicate that the method is most applicable above about 10 km. If one chooses wavelengths which allow contributions from very much lower altitudes, one also receives contributions from the surface which are not so easily evaluated. And worse still, one sees contributions from lowlying clouds, aerosols, etc., which are not taken into account in the calculations which refer to a clear atmosphere.

In fact, it is not possible to evaluate the effectiveness of the infrared method just described until extensive experiments have been undertaken. These must be carried out from at least balloon altitudes and should cover a variety of meteorological situations, as well as different latitudes and, ideally, different seasons.

In order to evaluate the method it would be essential to have an instrumentation which is much more elaborate than the one which is now being built for unmanned weather satellites. The latter observes at only six wavelengths but in a manned vehicle a more elaborate setup could be accommodated, covering the 15 micron band at many more wavelengths and covering other absorption bands as well for the sake of comparison. (It may be expected that the contribution of aerosols is different at 15 micron than it would be at other wavelengths since the size of the aerosol particles plays a determining role in their effectiveness for radiation emission.)

A very important question which can only be settled properly by direct experiments is the comparison of the CO_2 15 micron experiment with a corresponding one in the microwave region based on the absorption properties of molecular oxygen. In the microwave region, the disturbing effects of aerosols and even of clouds are much less pronounced; on the other hand, the effects of the surface become more uncertain.

It turns out that the technical realization of the microwave experiment is quite difficult. It requires very low noise receivers which cannot as yet be accommodated within the limited payload and power capability of unmanned weather satellites.

Presumably, however, in a manned space station this difficulty would not exist and, therefore, it would be possible to conduct a simultaneous comparison of the infrared and microwave experiments.

7. Water Vapor

The vertical distribution of water vapor can in principle be obtained by measuring either its infrared or microwave emission at a number of convenient wavelengths in the absorption bands of water vapor. The result gives the temperature corresponding to a certain pressure depth of water vapor. But if the temperature–altitude curve is independently measured as just discussed above, then one obtains the pressure of water vapor at various levels in the atmosphere and, therefore, its vertical distribution. Only very crude experiments have been carried out in Tiros satellites but they indicate that the method is feasible, at least in principle. Its ultimate accuracy, which will be short of what is possible theoretically, cannot be assessed again until actual experiments are carried out. In order to choose the proper wavelengths for observations, quite protracted experiments will be necessary, either from balloons or from manned satellites, involving an examination of all of the likely absorption bands of water vapor in the infrared and microwave regions to determine which one or which combination of bands is best suited to give the highest accuracy.

Again, a manned satellite vehicle has the great advantage of allowing perfectly controlled adjustment of the equipment.

8. Radiation and Heat Balance

The overall energy balance of the atmosphere can be obtained quite directly from satellites. The satellite can measure to any detail desired the radiation emitted from the earth below without distinction as to where it originates, whether at the surface, in the atmosphere, or from clouds, and it can also measure the total of the solar radiation which is scattered back by the atmosphere or reflected from clouds or from the surface. Such measurements have already been carried out, albeit crudely, from the Tiros satellite. However, it remains for meteorologists to devise methods for utilizing this information effectively in numerical predictions of long-term weather changes.

9. Surface Emissivity

Reference has been made from time to time to the infrared and microwave emission from the earth's surface. This emission is not at all well known. It clearly depends upon temperature and emissivity; the latter depends on the state of the surface, whether it is a land surface or an ocean surface, whether it is covered with snow or ice. (It even depends on the state of the ocean and, therefore, may give a useful way of measuring sea state.)

Indications are that the infrared emissivity of the surface is high but not quite equal to one and that it drops off towards lower values in the microwave region, especially for ocean surfaces which have very low emissivities. This means that in the microwave region an ocean surface will emit very little radiation and, therefore, appear to be quite cold to a microwave radiometer. On the other hand, the ocean surface reflects any radiation which falls upon it from above. In a wavelength region where the atmosphere is completely transparent and has no absorption bands and therefore no emission, the ocean surface simply reflects the microwave noise which is incident from extraterrestrial sources. However, as we move into the wings of the atmospheric absorption bands, the ocean surface will appear to warm up; and with a disturbed ocean surface where the atmosphere is viewed at larger zenith angles, the effective temperature would increase considerably. Therefore, it is possible in principle to measure the sea state by determining the emitted microwave power at certain wavelengths.

This is a passive method for measuring sea state. An active method depends on reflected solar radiation-sun glint. Sun glint has been noted in many satellite pictures and the size of the sun glint area gives a fairly direct indication of the smoothness of the ocean surface. Work is now underway to provide a more quantitative relationship.

In principle, the state of the ocean surface could also be investigated again by artificial light sources; i.e., by lasers. A smooth ocean surface will give specular reflection to an incident laser beam. A disturbed ocean surface will scatter the beam. It appears to be possible to devise a satellite experiment which can take advantage of this phenomenon, but the technique would have to be tested carefully.

The peculiar property of low emissivity of the ocean surface in the microwave region can also be used to detect and measure precipitation over the oceans. Essentially, as suggested by K. Buettner, the collection of water droplets will appear to be a warm spot superimposed over a 'cold' ocean surface. Again, many experiments will have to be undertaken before effective use can be made of this phenomenon; but in principle it appears possible. It would allow us to study the important process of energy exchange between the atmosphere and oceans which usually escapes detection.

The combination of infrared and microwave techniques can be exploited further for more detailed examination of the earth's surface. For example, it has been suggested repeatedly that the thickness and water content of snow cover could be estimated in this manner, as well as the thickness of sea ice and perhaps its closeness to breakup. Again, experiments are lacking; some of these can be carried out from platforms at rather low altitudes, from aircraft and from blimps, before a satellite experiment is designed. Unmanned weather satellites do not constitute the most efficient way of testing the basic parameters of such an experiment. On the other hand, manned satellites which carry elaborate infrared and microwave instrumentation can easily investigate these parameters by simply changing the wavelengths to a region in which the atmosphere does not absorb and in which the surface, therefore, becomes visible.

10. Active Radar Experiments

There has been a good deal of discussion, both pro and con, concerning the advantages of an active radar in weather satellites. The most common objective has been directed to the detection of droplets and precipitation – somewhat in the manner of a ground-based weather radar. However, active radar in satellites may be better suited to quasi-meteorological applications: determinations of sea state, snow and ice characteristics, and terrain properties. In distinction to passive microwaves (discussed above), one would be concerned here with measuring a reflection coefficient, and perhaps polarization, rather than thermal emission.

A manned experimenter would be in a good position to use the radar (or radars) selectively, directing it against specific ground targets, varying the power, pulse width, and especially the frequency, to take advantage of the prevalent meteorological situation.

11. Sferics

Thunderstorm detection by the emission of radio noise bursts ('sferics') seems to be suitable for satellites. Again, an experiment must be performed to arrive at an optimum setting of frequency, bandwidth, sensitivity, etc. – to detect what is best adaptable to meteorological use and to avoid man-made interference.

12. Analysis of Cloud Tops

One of the most important areas for measurement available to a satellite is the detailed investigation of the tops of clouds including the determination of cloud temperature, cloud altitude, and size distribution of cloud particles.

The cloud altitude in terms of pressure can be measured by using the cloud as a reflector and determining the thickness or depth of overlying atmosphere, according to the proposal of Hanel, Wark, and Yamamoto. Use can be made of the absorption bands of molecular oxygen, particularly those in the vicinity of 7600 Å. In the present proposal, the sun is used as a source. However, it may be entirely feasible to use a laser operating at the proper wavelength. In fact, since the wavelength of a laser can be varied under certain circumstances, it is possible to investigate the region in the band and near the band; the ratio between the two establishes the absorption due to the oxygen and, therefore, the cloud altitude (in terms of pressure).

In this determination, the assumption is made that the cloud is an adequate reflector. This point can be checked, however, by using a pulsed laser and determining the travel time. This direct measurement gives us cloud altitude in geometric terms; i.e., the actual distance of the cloud top from the satellite. It also allows us to establish the time spent by the light pulse in the cloud's interior before re-emerging. A further approach is the measurement of the infrared emission from the cloud top which is a measure of temperature.

The three measurements together thus establish the actual height in the atmosphere, and the temperature and pressure at that point. It is clear that such measurements are possible whenever there exists an adequate reflector; e.g., a cloud.

In principle this method could be used to measure the pressure at sea level, in the absence of clouds. However, it is very likely that the accuracy will be so poor as to make such measurements meaningless. Nevertheless it is important to carry on appropriate experiments. Because of the complicated absorption structure of oxygen, it will be necessary to carry out such experiments with a great deal of refinement of technique, again requiring bulky apparatus which is most suitable to a manned orbiting laboratory.

It is important to note that the cloud top measurements just described can be carried on at night if a laser is used, so that the method has quite general applicability.

The properties of the cloud particles can be investigated in detail by studying the reflection properties of sunlight or of laser-produced light. The reflectivity at different wavelengths can be related to a size distribution of particles. The important region here extends into the near infrared and as far out as perhaps 4 to 5 micron. As strong laser sources become available at longer wavelengths, such sounding techniques will become feasible, as indeed they are in the microwave region where weather radar is used to determine the existence of water droplets.

13. Ozone Distribution

Since ozone absorbs strongly in the region between 2500 and 3000 Å, it is possible to establish its distribution in the stratosphere by measuring the amount of back-scattered radiation in this wavelength region. Solar radiation has commonly been suggested. However, the method can be extended by using the moon as a source or even starlight, so that in principle a day and night method exists for measuring ozone distribution. No experiments of this type have as yet been carried out from a satellite although some rocket surveys of scattered ultraviolet radiation have been performed.

Again, it will be desirable to carry out the initial experiments with elaborate instrumentation having large light-gathering power and extremely high resolution, in order to establish the best operating conditions for an eventual unmanned experiment. Furthermore, there is likely to be interference to the ozone measurements arising from the existence of high altitude cirrus clouds. In order to establish their existence, one might want to use a laser detection technique so that a diagnostic service would be provided to explain any unusual results of the ozone instrument.

14. Conclusion

It is quite clear that what is needed in order to design suitable instrumentation for studying the atmosphere on a routine basis is a veritable meteorological observatory carrying at the same time all kinds of instruments having the highest possible resolution, light-gathering power, and flexibility in wavelength setting, etc., in order to provide data which give complementary information. This feature, more than anything else, may explain the immense usefulness of a large manned space station for meteorological research.

It is one of the characteristics of the atmosphere, which is a very complicated system, that one has to ask the same question in a number of different ways. One, therefore, has to build constructive redundancy into the instrumentation. An example is the temperature structure of the atmosphere. It may have to be measured in the 15 micron bands of carbon dioxide as well as in the microwave region absorption of molecular oxygen, or perhaps in other ways. The point is that all of these methods must be tested out simultaneously in order to be able to get a cross-check. There are different disturbing effects which enter in, for example, high cirrus clouds would affect the infrared but probably not the microwave. On the other hand, the contribution from emission from the surface of the earth will affect the microwave region in a more uncertain manner than the infrared. It is clear, therefore, that many of these problems can only be settled by direct test. A manned platform has the advantage of allowing a large weight and power capability and, therefore, makes it possible to fly several experiments at the same time. In addition, the presence of the man introduces great simplifications in the design of the equipment. It eliminates the need for many items which would have to be included in an unmanned experiment.

It proves the old statement: "Man is the lowest cost 150-pound non-linear all-purpose computing system which can be mass-produced by unskilled labor."

In justifying at least part of the manned program for atmospheric science studies one has to grant several assumptions. First and foremost is the assumption that we are willing to improve meteorological satellite instrumentation and are willing to spend money on this. This assumption can, of course, be defended on the grounds that the improved information about the atmosphere leads to the possibility of making improved forecasts of weather, of climate changes, and to the fundamental scientific basis for weather and climate modification. The economic value of all of these possibilities is well appreciated, although it has not been detailed sufficiently well.

At this point, it might be appropriate to consider some of the real benefits to be derived from such research. Economic estimates of improved weather forecasting have been made many times and these estimates predict savings of the order of billions of dollars to our economy arising from improved forecasting over a period of only a few days. Of more basic scientific importance is our understanding of the behavior of the atmosphere over longer periods and of understanding the causes for climatic changes.

15. Scientific Questions for the Future

Satellites give us a method of continuous surveillance of the earth's atmosphere. We can observe global changes with much greater sensitivity than any previous method at our disposal. We should be able, therefore, to study perturbations of the climate produced, for example, by volcanic eruptions which put a great amount of dust into the atmosphere. We should be able to study perturbations produced by large solar outbursts which have pronounced effects on the upper atmosphere and perhaps some effects on the lower atmosphere as well.

It is generally believed that the ozone is a more sensitive indicator to solar effects than any other meteorological parameter. To what extent is the ozone distribution responsible for climatic changes and to what is it a consequence of climatic changes? To what extent is the climate controlled by ocean temperatures and ocean currents, by global snow cover, by the extent of the ice cap in the polar regions, and by long-term variations of world cloudiness?

We may even be able to answer some rather fundamental questions. For example, is the present climate of the earth the only one possible considering the earth's environment, composition of atmosphere, state of surface? What happens when external factors perturb the climate? Does it return to its original basc valuc? Are there discrete levels in the climate? Or is it possible to speak about a continuously varying climate?

The answers will not be known for some time but a beginning is being made in our search to understand one of the most important problems facing man in his environment on this planet.

Bibliography

For a concise discussion of the four basic approaches to weather prediction (analog, statistical, physical, numerical) see; e.g., M. Tepper, 'Meteorological Satellites' in *Space Exploration*, McGraw-Hill, New York, 1964.

A detailed discussion of problems relating to meteorological satellites is given in the April 1963 issue of *Astronautics and Aerospace Engineering*. See also the book by K. Y. Kondratiev *Meteorological Satellites* (in Russian), available in English as NASA Technical Translation TT-F177 (1964) from OTS, Washington.

Atmospheric optics and radiation are discussed in the February 1964 issue of *Applied Optics*, and in the book by R. M. Goody *Atmospheric Radiation* (Oxford, 1964).

With regard to the technical details of remote sensing, consult *Proceedings of the Symposium on Electromagnetic Sensing of the Earth from Satellites* edited by R. Zirkind, Polytechnic Institute of Brooklyn Press, 1967, and especially the series on *Remote Sensing of the Environment* published by the Institute of Science and Technology, University of Michigan, Ann Arbor, Mich.; the Proceedings of the Fifth Symposium appeared in 1968.

TERRAIN SPECTRAL IMAGERY FROM SPACE IN VARIOUS SPECTRAL REGIONS

K. YA. KONDRATYEV, V. D. STEPANENKO, and B. V. VINOGRADOV
University of Leningrad, U.S.S.R.

Abstract. The present review discusses the possibilities and results of interpretations of earth observations from satellites for use in geosciences and other investigations of natural resources.

Different regions of the electromagnetic spectrum can give different and often complementary information about the atmosphere, the surface (land, vegetation, ocean, ice, hydrology) and subsurface.

We discuss:

(1) Space photography, black-and-white, as well as multi-spectral, with application to terrestrial natural resource problems.

(2) Infrared pictures of the earth in transparency windows and absorption bands, with possibilities of their meteorological interpretation.

(3) Space spectrophotometry; the identification of the spectra of surface formations with the purpose of discerning the mineralogical composition of rocks and for geobotanical investigations.

(4) 'Radar pictures' of surface formations; their advantages and the possibilities of active and passive microwave sensing.

1. Introduction

Television pictures of cloud cover from meteorological satellites have now become a necessary ingredient of operationally-obtained meteorological information which is the basis of weather forecasts. This is also true of *infrared* cloud cover pictures. An important peculiarity of infrared pictures lies in the fact that they can be analyzed not only qualitatively (judgments on cloud spatial distribution, structure and forms) but also quantitatively (determinations of the earth's surface temperature or cloud upper boundary as well as determinations of cloud cover height).

While the analysis of cloud cover pictures is mainly interesting from the point of view of meteorological applications, terrain images can also be used for the many aspects of geoscience, as well as for geobotanical, agrobiological and other such investigations. For the latter the peculiarities of the pictures, such as their high spatial resolution and multi-coloredness, become extremely important. Multi-coloredness ('multi-spectral') here implies the possibility of obtaining color photographs, spectra of surface formations, and direct or transformed pictures of the earth's surface in the various spectral regions, from the visible up to the microwave region. In this respect, the deliberate choice of objects to be investigated plays an important role, particularly because, in many cases, obtaining images of the earth's surface is hindered by cloudiness. These facts, plus the considerable complexity of the equipment involved, determine the need for using manned orbital laboratories for such investigations.

The present review is devoted to a discussion of the possibilities and results of interpretation of earth pictures as observed from space for use in the geosciences and

S. F. Singer (ed.), Manned Laboratories in Space.

other investigations of natural resources.* The data are considered from the point of view of the spectral region of electromagnetic waves, depending on the form they are received on board the space vehicle (photographic registering of the visible light and near infrared radiation field of the earth, spectrometric, radar and microwave surveys).

2. Space Photography

When compared with all other methods of registering the electromagnetic radiation field of the earth, the largest quantity of information presently can be obtained by photography in the visible and near infrared spectral regions, in wavelengths ranging from $\lambda = 300$ nm** up to $\lambda = 1100$ nm. In practice, however, a range of photoactinic fluxes from $\lambda = 550$ nm to $\lambda = 850$ nm is used. This refers to both original photographs, delivered to the earth directly in containers, and to television pictures, although the former are much to be preferred. Characteristics of photographs of the earth taken from space are their high resolution, the satisfactory contrasts in photographic imagery of natural objects, the detailed reproduction of the elements in the landscape, and the rather good distinction in the color or tone of the elements. The potential possibilities for improvement in space photography are also important [10, 26, 46, 66, 68].

The main geographic advantages of space photography for the study of natural resources are reduced to the integration of natural characteristics in three directions: (1) territorial integration; (2) factorial integration; and (3) dynamic integration [78].

Territorial integration implies the combination of separate morphological elements of the geosphere into large territorial complexes by means of direct photographic generalization, over a small scale, of pictures of the earth's surface. An advantage of space photography here is the possibility of identifying large territorial complexes and their limitations by tracing directly from small-scale photographs, thus escaping the multi-stage cartographic generalization necessary from large-scale field survey maps to small-scale maps of 1:1 000 000 or smaller. This permits interpolation, and the inevitable errors connected with it, to be replaced by direct observation and extrapolation. The use of space photographs offers the investigator a method for surveying and evaluating natural resources on a planetary scale.

A first attempt at territorial integration has been the compilation of a planetary scheme of cloud distribution over a 24-hour period [1, 9]. Territorial integration appears particularly promising for regional investigations in the areal fault zones, planetary jointing, large tectonic structures, and geographical zonality. Practical experience in the application of territorial integration to space photographs for the investigation of natural resources is rather restricted, however. From photographs obtained by the Viking rockets, the regularities of the geological and geomorphological structure of the State of New Mexico, U.S.A. [49] have been studied. From photo-

* Since a vast amount of literature, including review papers, is devoted to problems of meteorological interpretation [34, 35, 36, 37, 39], a number of related problems will not be discussed here.

** 1 nm (nanometer) $= 10^{-9}$ meter $= 10$ Å $= 10^{-3}$ micron.

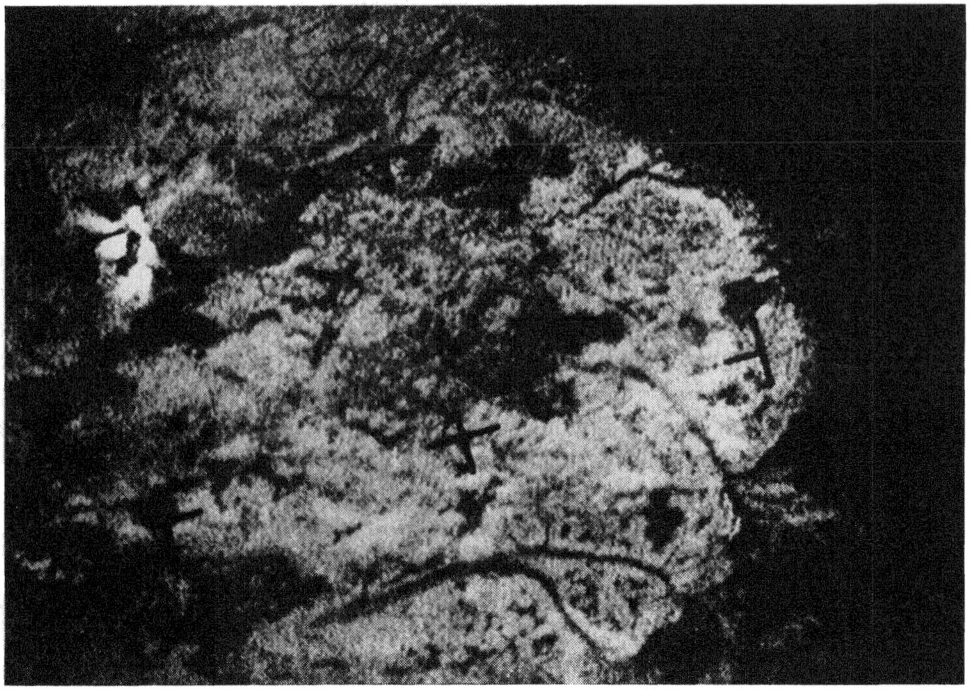

Fig. 1a. The television-picture of the Paris Basin obtained from the Nimbus-1 satellite on September 14, 1964 (improved high resolution television-camera); at the scale of about 1:2000000.

graphs taken by the MA-4 space vehicles, geological maps at a scale of 1:2 000 000, have been made of the Northwestern Sahara and details (faults, flexures, separate outcrops of strata) not known before have been discovered [56]. The use of territorial integration can be illustrated by the experimental compilation of a landscape map and regionalization of the Paris Basin from a 1:2 000 000 scale picture obtained from the Nimbus satellite [79]. This latter picture allows one to clearly distinguish seven landscape types and regions (Figures 1a, b).

Factorial integration is based upon the combined registration on one space photograph of both landscape physiognomy components and latent ones, discernable by means of natural indicators. Factorial integration consists of the simultaneous interpretation of various components of the geosphere: those of the climatic sphere, hydrosphere, biosphere, and anthroposphere, and their integration according to intercorrelations into definite natural complexes. The use of space photographs has permitted, for the first time by photo interpretation processes, a complete integration of all the components of the geosphere; from the deep geological structure to the upper atmospheric layers, including meteorological conditions and a number of recognizable landscape components. The interpretation of the relations between phenomena from space photographs, apart from simultaneous identification of landscape components in the same area, is also done by tracing the forms of interconnec-

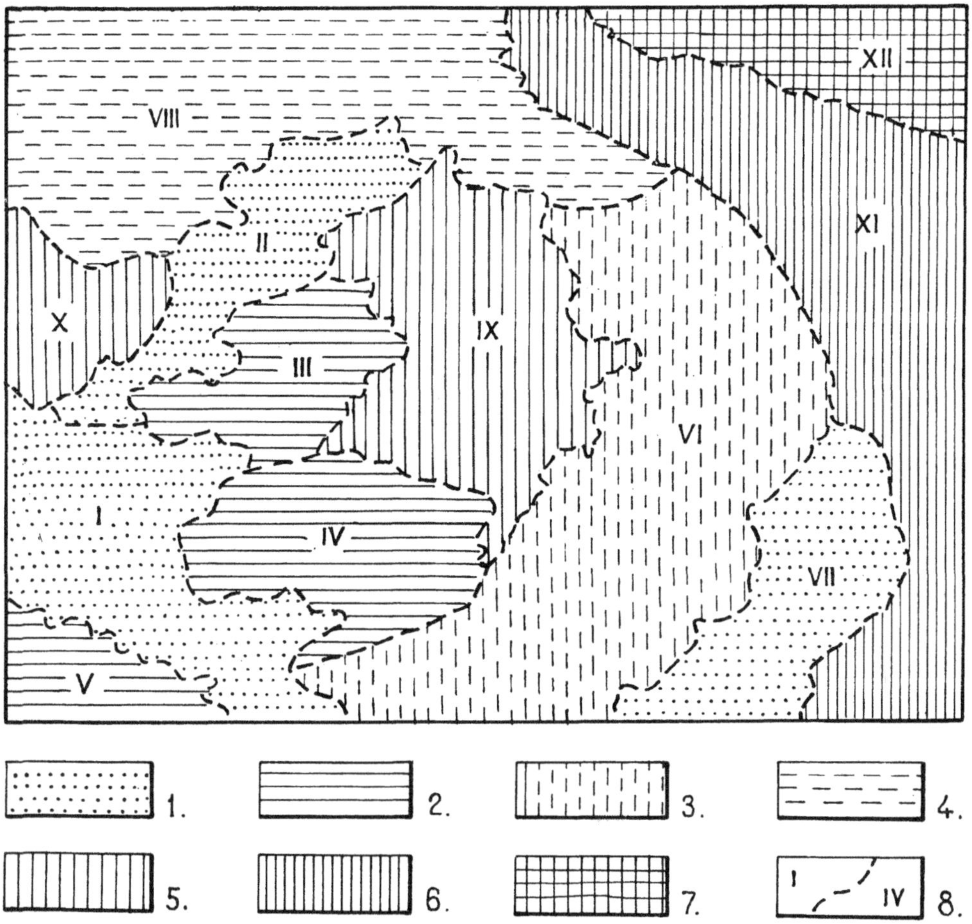

Fig. 1b. Territorial integration: the landscape subdivision of the Paris Basin at the scale of 1:2000000 from the picture obtained with the Nimbus-1 satellite (Figure 1a). – (1) The landscape of the lowlands of the Seine and Oise Rivers is represented by the winding lines of the big rivers, the dark compact spots of oak-hornbeam forests on the hillsides, and the uneven grey to light-grey tone of the arable land. – (2) The landscape of the flat limestone plateau of Brie, Valois, Beauce (mostly treeless) is characterized by fields of even light-grey tone divided by dark-grey elongated stripes of oak forests situated along the cuestas, tectonic lines, and river valleys. – (3) The landscape of the low chalky cuesta ridges of Champagne (mostly wasteland) is distinguished by the lightest light-grey tone, together with the low-contrast large marbly pattern of summer oak-pine forsests which is extremely noticeable against the background of narrow strips of transit rivers. – (4) The landscape of the seaside plains of Picardie is notable by the homogeneous light-grey to grey tone with the low-contrast fine texture and dark winding lines of the river valleys. – (5) The landscape of the hilly elevations of Marne, Bray, with its oak forests, is represented by a vaguely-spotty, relatively-contrasting pattern consisting of compact, dark-grey to darkish-grey spots of forests, occupying the largest part of the entire area and separated by narrow light-grey spaces of wasteland. – (6) The landscape of the cuesta elevations of Argonne is identified by length-wise-oriented, dark to dark-grey spots and the stripes of oak-beech forests on the ridges, separated by the strips of consequent valleys with arable lands. – (7) The landscape of the mid-high mountains of Ardennes, covered by oak-beech forests with an admixture of birch and spruce, can be distinguished partly by the massive, elongated dark spots of the forests and partly by the low-contrast texture of the open woodlands.

tion between those and other landscape components, particularly boundary phenomena, and subsequent comparison of genetically-related, areally-associated landscape components.

For instance, the connection between the relief and composition of the bearing surface, and the forms and distribution of cloud formations has been noted repeatedly: the foehn situation in windless regions of elevated islands and mountains; the connection between the presence of convective cloud forms and orographic barriers; the influence of sea coast lines on the distribution of stratus clouds, fog, cumulus clouds; the connection between parallel undulated cumulus clouds and jet stream boundaries of warm water masses, etc. [8, 11, 23]. The interconnections between other landscape components can be traced just as distinctly, although these have not been subject to special investigation and so are seldom mentioned. The connection between the composition and distribution of vegetation, soils and arable land with the geological structure, composition of rocks, relief and hydrological conditions are the most evident. Thus, for instance, the pre-Cambrian intrusive mass of Ifni (Morocco), which is composed of rhyolites and dactites in dikes and stocks, andesites and granites, and covered by xeromorphic open woodlands with argania and scrub oaks, is represented by compact dark-grey spots with a low-contrast, irregularly-rounded, spotty inner texture connected with the vegetation distribution at the limits of the body of intrusion.

The chalks of the Upper Cretaceous of Champagne (France) are composed mainly of wasteland, rare scattered summer oak forests, and pine plantings. They stand out among the surrounding forest territories due to their light to light-grey tone and low-contrast pattern. A comparison of the areas of hydrophytic vegetation in the deep deflation basins of White Sands (New Mexico, U.S.A.) with the direction of the temporary run-off and inclined fan belt of the Sacramento Mountains, and their absorption zones at the peripheries of large sand masses, shows a close relationship between all the above mentioned phenomena. The negative correlation between the distribution of arable land and masses of effusive rocks can easily be seen in the space photographs of the coast of the Gulf of Tonkin (Southwest China). All these examples illustrate the results of simultaneous photo interpretation of separate landscape components. The most complete form of factorial integration is the revelation of genetically-restricted natural complexes combining all landscape components.

The existence of orbital space stations makes it possible to obtain photographic pictures of any illuminated portion of the earth at different time intervals, from several minutes and hours to months and years. A comparison of repeated photographs will reveal the dynamics and rhythmics of landscapes. If repeated surveys are performed with the same photographic system, a high degree of comparability is possible. Thus, dynamic integration involves the recognition of temporal complexes, integrating the cycles of rhythmic and subsequent dynamic changes in landscape components. Successive photography will permit the dynamics principle to be brought into all forms of natural resources mapping. Whereas examples of dynamic integration in the investigation of meteorological situations and ice conditions are numerous and well-known and may therefore be ignored in the present review, the problems of investigating the rhythmics and dynamics of other landscape components from repeated space photographs have barely been developed.

The most reliable method of dynamic integration consists of the comparison of subsequent satellite photographs of the same terrain received by the same recording system. Investigations of the distribution of snow-covered areas and of the rates of snow-cover formation and thawing, from repeated surveys, were performed in the Sierra Nevada Mountains and Adirondack Mountains (U.S.A.), Southeast Canada, and the Alps [13, 20]. Experience in the interpretation of vegetation phenology from successive photographs has not been sufficient nor convincing enough to date [13, 27]. Nevertheless, this use of space photography appears to be one of the most promising from the point of view of annual investigation of pastures in arid territories of the globe.

An indirect but no less important method of dynamic integration from space photography data is the ability to distinguish traces left on the landscape by periodic and catastrophic phenomena. Some examples are the detection of forest fires from smoke [6, 7] and determinations of rainfall and soil moisture variations from patterns observed [28].

One of the main factors to be considered in the interpretation of space photographs should be total integration. This assumes both simultaneous and associative identification of both geosphere components (atmospheric data, oceanographic phenomena, hydrological conditions, geological structures, geomorphologic elements and topography, vegetation, soil cover, some areas of zoocoenosis, land use, etc.) and territorial complexes of different ranges combining separate components (sites, land units, land systems, landscapes, etc.)

Ignoring the meteorological application of space photography, we note only a few significant papers, summarizing and characterizing the main uses of photographic information obtained from satellites for the investigation of atmospheric physics and meteorological situations [22, 34, 36]. A number of investigations devoted to meteorological interpretation are of methodological value and can be combined with the study of natural resources and other landscape components [1, 5, 9].

Space photographs have proved very successful for hydrological and oceanological research including investigations of ice cover distribution and dynamics of the oceans, seas and lakes [75, 80], as well as in studies of sea currents, temperature distribution of water surface, disturbance [79, 60], snow cover distribution and dynamics [13, 21] and ice sheets and glaciers [62].

Experiments in geological and geomorphological interpretation of space photographs have been quite successful. A number of papers offer examples of the study of rocks, geological structures, tectonic displacements, genetic types of the reliefs of North America [47], the Western Sahara [57], and South Africa [33]. Successful investigations are being carried out in an analysis of the geological structure of Arabia. The EROS satellite program is expected to provide much information for the purposes of a geological survey.

The least-developed interpretations, and those seldom included in research programs, are of vegetation, soils and agricultural lands. However, it is apparent that the physiognomy of these landscape components creates wide possibilities for their investigation from space photographs. The paper by Morrison and Bird [55] gives the first examples of the ability to distinguish vegetation. A few investigations have been devoted to the study of forests [82] and various kinds of agricultural land [10]. Only the interpretation of humidity has thus far attracted attention in the investigation of soils [28]. Much is expected of a spectral multi-zonal space survey in the study of vegetation and soils [7, 43].

The limitations of a brief review make it impossible to discuss all aspects of the use of space photographs for the study of natural resources. The ideas expressed, however, are suitable for every type of investigation.

A number of methodological and technical problems must be resolved before the use of space photographs for the study and development of natural resources can be expanded.

(1) The regularities of photographic reproduction of natural objects from space must be more carefully studied. These regularities are determined by the spectral reflectivity coefficients of the different objects, the structure of natural formations, and the general and spectral transfer functions of the atmosphere [10, 19].

The optimum technical means and natural conditions for surveying from space should be determined by theoretical investigations and experiments, involving such optimum parameters as characteristics of films, color filters, cameras, as well as scales, seasons, weather, time of day for each different geographical zone and landscape of the earth [24].

(2) The methodology of interpretation of space photographs of the earth is based, as a rule, on the conclusions made in the analysis of the aerial photographs: the study of key sites, the compilation of photo interpretation indices and samples, the systematization of photographic information. The typology of television pictures of cloud structures is an example of the compilation of photo interpretation samples, graphical and logical keys. A system of key sites, characterizing the main landscape types of the earth, combined with aerial photography and aircraft observations of the atmospheric state, is the necessary scientific foundation for systematic development of a methodology for interpretation of space photographs of the earth [4].

(3) A technology for automating the processing and analysis of the large volume of photographic and geographic information received will have to be developed. Of great interest in this respect is the automatic identification of cloud structure types from photometric features; the brightness or density means, dispersion, autocorrelation functions, etc. [5, 65, 71]. A quantitative analysis of the distribution of photographic densities will allow one to objectively choose the natural and technical parameters needed for surveying and to determine the minimum samplings of key sites, photographic samples and test measurements. Finally, the development of a means for automatic mapping and storage of geographical information is of vital importance [63].

3. Space Spectrophotometry

The highly-informative content of black-and-white, color, and multi-spectral photographs of the earth's surface indicates that information about the spectral peculiarities of separate regions of the earth's surface should prove of great value in supplementing the mainly qualitative analysis possible through the use of monochromatic pictures, with quantitative evaluations based on spectrophotometric data.

Various aircraft methods for investigating the properties and characteristics of different regions of the earth's surface have been developed using reflected solar radiation spectra. Even more promising are similar investigations made by the analysis of infrared, thermal and microwave radiation spectra. The use of the latter is more advantageous, since neither atmospheric haze nor clouds normally prevent the registering of microwave radiation.

Similar methods can be applied to satellite use. Although the influence of the thickness of the atmosphere becomes a factor which complicates satellite data interpretation, the advantages of a global information scale, with deliberate choice of objects to be investigated, make it both desirable and well-suited to manned spacecraft.

The amount of information possible through the use of a manned spacecraft is very great, so the experiment can be multi-purpose. Some of the possible investigations might include:

VISIBLE AND NEAR INFRARED SPECTRAL REGIONS

Investigations of the spectral reflectivity of natural objects illustrate the great diversity in albedo spectral characteristics, as well as in brightness coefficients. Despite this, it is possible to divide all the surfaces into the following groups [40]: (1) soils; (2) kinds of vegetation; (3) snowy and icy surfaces; (4) water surfaces. Cloud formations may also be added to these groups. Figure 2 shows the albedo spectral curves which are characteristic of the above groups [38, 18].

The albedo of the cloudiness is given from calculations referring to stratus water clouds (the cloud optical thickness $\tau=30$, the solar altitude $h=60°$). The albedo values given for the wavelengths of 550 and 700 nm are the mean ones for stratus and strato-cumulus clouds with $h=60°$. It should be noted that, according to the calculations, the albedo of clouds in the visible spectral region does not depend on the wavelength [18].

Figure 2 clearly illustrates the specific character of the spectral reflectivity of various natural formations.

Knowledge of the spectral characteristics of the albedo of various surfaces permits one to try to solve the inverse problem; i.e., surface identification from spectral reflectivity data. On the whole, the spectrum is naturally most informative. It is possible, however, to find a simpler solution by considering the albedo ratio for two wavelengths.

For satellite data, the solution to the problem may be sought by the investigation of the albedo, for instance, for the two spectral regions of 650 and 850 nm and by the determination of their ratio K. The spectral regions of 650 and 850 nm were selected

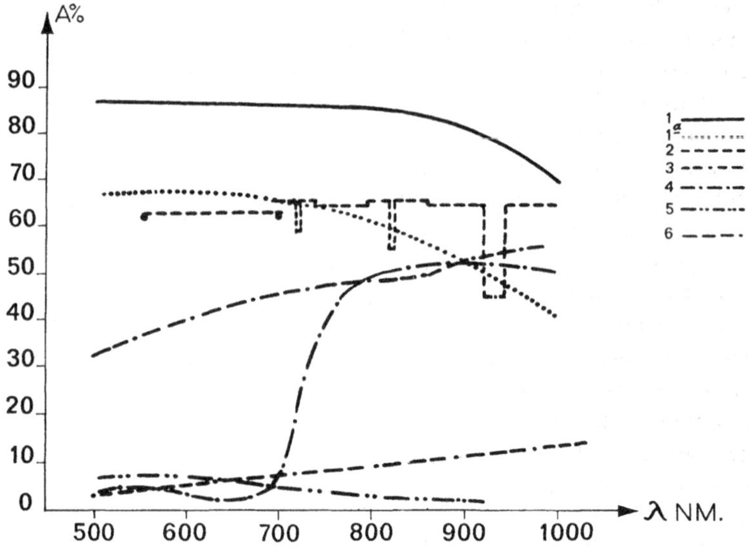

Fig. 2. The spectral albedo of natural underlying surfaces and clouds. – (1) Snowy surface (fresh-fallen snow), the solar altitude $h_{\odot} = 30°$. – (1a) Snowy surface (wet snow), $h_{\odot} = 30°$. – (2) Clouds (stratus), $h_{\odot} = 60°$, the cloud optical thickness $\tau_0 = 30°$. – (3) The sand (white), $h_{\odot} = 47°$. – (4) Lucerne (dense grass stand), $h_{\odot} = 52°$. – (5) Water surface (the lake at the depth of 70 cm), $h_{\odot} = 58°$. – (6) Black earth, $h_{\odot} = 48°$.

because of (1) the maximum transparency of the atmospheric thickness for these wavelengths and (2) the highest 'sensitivity' of the K coefficient to the type of natural formation. The use of surface observation data [38] and calculated data for the cloud albedo [18] leads to the following values for K:

surfaces covered with vegetation:	4–15
soils:	1.2–3.0
snowy surfaces:	0.75–1.0
water surfaces and clouds (water clouds):	0.9–1.0

As can be seen, only one albedo ratio for two wavelengths is a rather ambiguous characteristic of a natural formation. This means that more complete and complex factors need to be sought. (It is clear, for instance, that the use of an absolute value of albedo as an additional characteristic in the visible spectral region will result in a more definite situation.) Only green vegetation cover stands distinctly apart due to its high K coefficient value which is caused by the abrupt increase of reflectivity in the near infrared spectral region.

Figure 3 shows the spectral distribution of the relative brightness of wheat and oats in the visible and near infrared spectral regions, according to the data of Low and Shay [44]. The vertical dimensions of the circles and rectangles show the range of brightness variation. This example clearly illustrates the advantages of vegetation cover identification from reflection spectra, made sufficiently unambiguous due to the amount of information available.

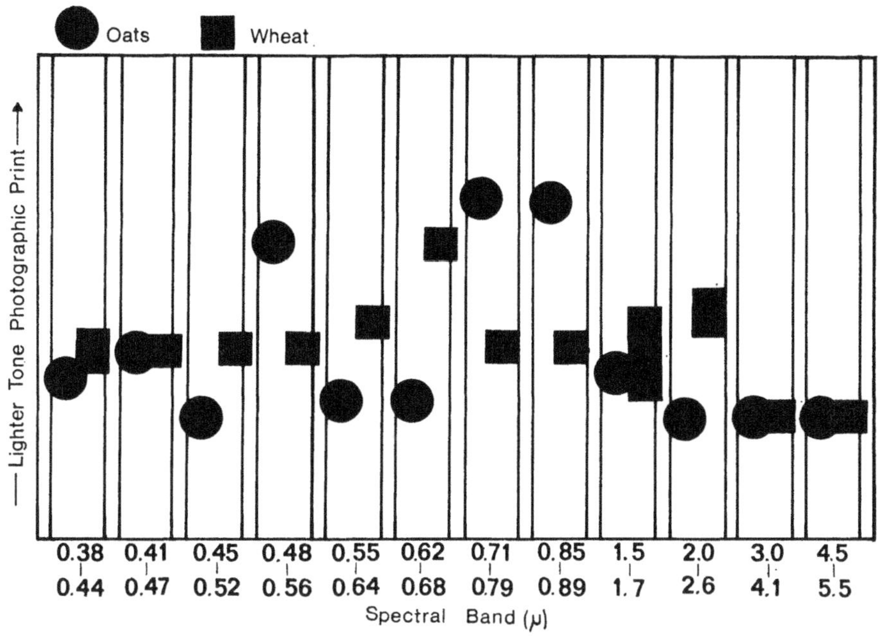

Fig. 3. The spectral distribution of the relative brightness of wheat and oats in the visible and near infrared spectral regions.

Investigations of the state of crops, of forest vegetation, of harvest estimates, etc., are possible due to the high sensitivity of spectral 'images' on the state of vegetation cover. The success of these investigations, however, is evidently impossible without preliminary modeling and surface tests. The problem of soil investigation may be solved in a similar way.

The most complicated problem is that of distinguishing snow from cloudiness. The wide range of variability of cloud albedo and, in a number of instances, in its identity with snow albedo (snow also giving great reflectivity variations), results from the fact that clouds can be water-droplets or ice, and from the variable thickness of the cloud layer. Analysis of measurements conducted above clouds and snowy surfaces shows that additional data on the radiation reflection in the wavelength of 1.6 μ are necessary.

In the wavelength interval of 0.4–1.3 μ the brightness of water clouds is somewhat smaller than that of snow [77]. This also holds true for the albedo, according to calculations provided by Avaste [2].

In the 1.4–2.5 μ wavelength interval, cloudiness is brighter than snow, the contrast reaching 65 to 75% for wavelengths of about 1.6 μ [77]. This contrast allows one to distinguish a snowy surface from a water cloud. Thick ice clouds have reflectivity spectral characteristics similar to those of snow and are scarcely distinguishable, therefore, from snow.

It should be stressed that a successful solution to the problems discussed is made more difficult by the extremely fragmentary data available on spectral reflectivity of natural formations.

4. Infrared Spectral Region

It has been known for a long time [6] that the spectra of reflection and self-emission (thermal radiation) are specific for different natural surfaces, depending particularly on the mineralogical composition. This is particularly true with respect to thermal radiation spectra. Since the development of high sensitivity receivers of radiation in sufficiently long-wave spectral regions (about 5 to 10 μ) has only recently been accomplished, thermal radiation spectra of minerals in the near infrared region (1 to 5 μ) have been investigated in more detail. The results appear to show that it is possible to identify minerals from their thermal radiation spectra.

Using data from the Croning and Clifford paper [14], Figure 4 illustrates the spectra of thermal radiation from polished samples of various rocks normal to the surface: quartzite, feldspar, dunite and calcite. The shaded areas correspond to the ranges of radiation absorption by the atmosphere. As can be seen, the spectral distribution curves of radiation intensity differ notably. Quartzite is shown by the presence of its minimums at wavelengths of 8.4 and 9.2 μ; feldspar at 8.6, 9.5 and about 10 μ; and dunite and calcite at 11.2 and 11.4 μ respectively. The typical minimum radiation of calcite at about 6.7 μ occurs in the zone of total absorption of radiation by the atmosphere. It should be noted, however, that, even in the atmospheric transparency window of 8–12 μ (schematically marked in Figure 4 as free from absorption), considerable selective absorption exists, particularly in the 9.6 μ ozone band.

The spectral peculiarities of thermal radiation depend entirely upon the structure (roughness) of the surface. For instance, if a sample is granulated, its spectrum corresponds to the radiation of a grey body; i.e., it loses its special qualities.

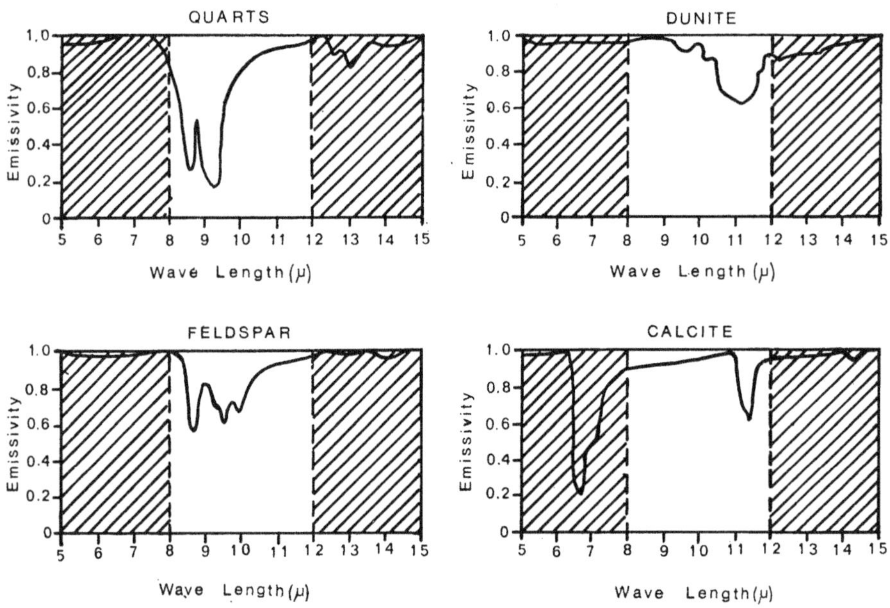

Fig. 4. The relative emissivity of polished samples normal to the surface.

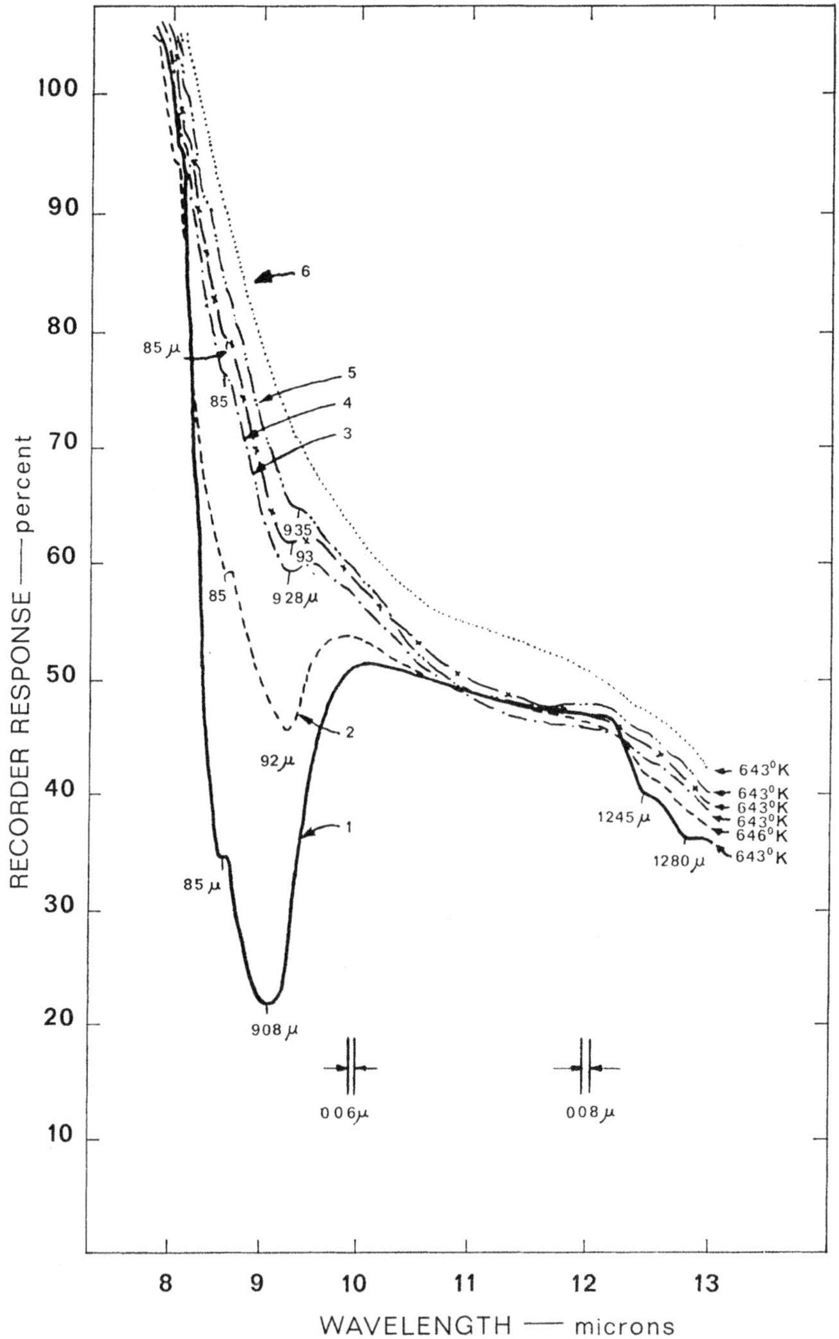

Fig. 5. The relative emissivity of quartz in dependence on the surface state and the degree of granulation (the emitting surface temperature was 643 ± 3 K). – (1) Polished surface (integral emissivity $\varepsilon = 0.73$). – (2) Rough surface ($\varepsilon = 0.81$). – (3) The size of particles – from 25 to 45 ($\varepsilon = 0.88$). – (4) The size of particles – from 10 to 25 ($\varepsilon = 0.90$). – (5) The size of particles – from 1 to 10 ($\varepsilon = 0.92$). – (6) Black-body radiation at the temperature of the air.

Figure 5 shows the radiation spectra of quartz depending on the state of its surface and the degree of its granulation. The individual features of the spectrum of a sample become obliterated by further granulation. Only the scarcely noticeable minimum of about 9.35 μ remains after granulation into very small pieces. It has been found possible to identify granite and dunite from radiation spectra only when the size of the particles is greater than 100 μ [6]. A similar criterion can be applied to many other minerals.

Taking into consideration the possible identification of minerals from their thermal radiation spectra, Burns and Lyon [6] have suggested an idea for a mineralogical satellite. Since high spatial resolution is extremely important in this case, it would be desirable to use a satellite in a low-altitude orbit.

Lowe and Coleman [45] have found that the only promising instrument for such investigations would be interferometer spectrometers which would allow almost instantaneous registering of the radiation spectra of low-temperature sources.

5. Space Radar and Microwave Survey

The great achievements in developing thermal microwave emission and radar, allowing information to be obtained at any time of day and almost independently of weather conditions, determine the possibility of using a wavelength range from 1 mm to 30 cm. Radar and thermal emission methods are widely used in surface and aircraft geophysical investigations [58, 57, 73, 16, 78, 17, 51]. If the above-mentioned equipment were used in satellites, there would be no major limitations for the problems in question. It would be necessary, of course, to overcome a number of technical difficulties connected with specific conditions of work in space, as well as with the energy, geometric resolution and insurance of equipment reliability and durability necessary to obtain satisfactory information.

Application of such equipment in the following stages of application would seem to be in order: first, the use of thermal microwave emission; second, the use of radar stations; and, third, the simultaneous use of both.

6. The Use of Thermal (Passive) Microwave Survey

The intensity of thermally emitted electromagnetic radiation received by a microwave receiver and expressed in terms of radiobrightness temperature is:

$$\begin{aligned} T_b = {} & T(1 - R^2)\exp\left(-\sum_{i=1}^{n}\tau_i\right) + T_1[1 - \exp(-\tau_1)] \\ & \times\left[\exp\left(-\sum_{i=2}^{n}\tau_i\right) + R^2\exp\left(-\sum_{i=1}^{n}\tau_i\right)\right] + \bar{T}_2[1 - \exp(-\tau_2)] \\ & \times\left[\exp\left(-\sum_{i=3}^{n}\tau_i\right) + R^2\exp(-2\tau_1)\exp\left(-\sum_{i=3}^{n}\tau_i\right)\right] \\ & + \bar{T}_n[1 - \exp(\tau_n)]\left[1 + R^2\exp\left(-2\sum_{i=1}^{n}\tau_i\right)\exp(-\tau_n)\right], \end{aligned} \tag{1}$$

where T is the absolute temperature of the underlying surface, R^2 is its reflection coefficient, τ_i is the optical thickness of the i-atmospheric layer, and $\bar{T}_i$ – the mean temperature of the layer.

Formula (1) is approximate because it ignores the degree of roughness of the surface, the effect of scattering by precipitation droplets on the radiation transfer, and assumes the homogeneity of properties within the thickness of the i-atmospheric layer.

The effect of scattering by non-rainy clouds can be ignored over the whole microwave range, for slight and moderate rains at $\lambda \geqslant 1.6$ cm, and for heavy ones – at $\lambda > 2$ cm.

Formula (1) becomes considerably simpler in the absence of clouds and precipitation. With the help of this formula, T_b and the ΔT_b contrasts can be quantitatively evaluated, the latter, in their turn, allowing to judge about energy contrasts of various objects on the earth and in the atmosphere.

Areas of land, due to their smaller, as a rule, reflection coefficients than those of water, give considerable radio-thermal contrasts (100–150° K) which allows one to detect easily the shores of continents, islands, rivers, etc. Experimental data show that from the ΔT_b values in the centimetric range clays, sands, turfs, etc. can be distinctly identified. This gives a chance of using thermal microwaves for the geological mapping of surface deposits [57]. The application of thermal emission microwaves may prove useful in the investigation of deeply frozen rocks and at the search of hydro-thermal waters.

Due to the relatively large wavelengths this method is 'deeper' compared with the use of infrared radiation, since the penetration ability of radio waves is measured by decimeters and meters. This makes it possible to judge about the composition and properties of rocks from absorption spectra [57].

The application of thermal microwave to the purposes of ice reconnaissance and iceberg detection has also proved very effective, since the reflection coefficients of water within the limits of sighting angles from 90° to 30° are notably higher than those of ice upon sea water [3]. This fact, together with the difference in the absolute temperature between water surface and ice, cause considerably larger values of ΔT_b than the minimal detectable ΔT_{min} values.

The use of a two-frequency thermal radar for the measurement of water content of snow seems promising as well [31].

The results of the investigation of the possibility of precipitation zone detection appear to be hopeful. The calculations based upon formula (1) in paper [73] show that the ΔT_b value depends upon λ, polarization, sighting angle, rain intensity, the temperature and properties of the underlying surface. For rains with the intensity $I = 1$–20 mm per hour falling over water, ΔT_b at $\lambda = 0.8$ cm are within the limits of 67–100 K with the sighting angle $\delta = 60°$ and the horizontal polarization, while for $\lambda = 3$ cm in the similar conditions $\Delta T_b = 8$–102 K. The values of ΔT_b for clouds without precipitation are of only several degrees. The ΔT_b contrasts of clouds and precipitation are much smaller when observed from land due to land's greater emissivity [58, 73].*

* This forms the basis for a method of measuring precipitation over the world's oceans [69,74].

The experimental data on the whole confirm the results of calculations. Aircraft investigations of a two-channel thermal radar at $\lambda = 1.35$ cm and $\lambda = 1.9$ cm have shown that the mean value of radiation coefficients for land and ocean at the sighting angle of 44.5° equals respectively 0.85 and 0.53. Areas covered with forest were approximately 5° colder than the surrounding woodless territory, and the signal from broken cumuliform clouds 700 m and 1400 m thick scarcely differed from that reflected from land [29].

The use of thermal microwaves also requires a satisfactory geometric resolution which can be achieved by means of air-inflated and collapsible antennae. The antenna, for instance, of a 4-mm thermal radar is assumed to have a 6-m aperture [58]. The above antenna should ensure a spatial resolution of the order of 300 m from the height of 300 km, with the time constant of 0.5 sec. A high resolution is desirable also because due to reflection non-coherence the obtained picture of the area should be similar to the visual representation.

The use of thermal radar working on $\lambda = 0.5$ cm and 1.35 cm, and on other wavelengths corresponding to the resonance absorption in oxygen and water vapor offers an opportunity for the determination of the vretical profiles of temperature and humidity, as well as of the general content of water vapor and oxygen in the atmosphere on the basis of using the interpretation method developed for the optical wave range.

With the temperature sounding from satellites on $\lambda = 0.5$ cm it is necessary to take into consideration the effect of Zeeman splitting of spectral lines caused by the magnetic field of the earth [41]. Due to it, the intensity of absorption becomes dependent on the angle and polarization. An approximate solution of the problem can be obtained from the data on the brightness temperature matrix. The use of several intensive absorption lines allows one to determine the temperature profile up to heights of about 75 km. The problem, however, is complicated by the fact of the dependence of the radio-brightness temperature matrices upon the geographical coordinates of the thermal microwave receiver mounted on the satellite, as well as upon the polarization and temperature stratification. Besides, one requires a high-sensitivity receiver, high accuracy and frequency stability.

The surface experimental investigations of the atmospheric absorption spectrum at five frequencies in the $\lambda = 0.9$–1.4 cm interval have shown that the form of spectral lines is qualitatively connected with the humidity vertical profile, and the transmission at the given frequency is representative of the general content of water vapor in the layer about 15 km thick [72]. The above mentioned absorption spectrum is similar to the atmospheric radiation spectrum registered by the satellite equipment provided it is over the ocean (it is essential that the reflectivity of the water surface is high, as well as the fact that atmospheric transmission for the wavelength about 1 cm is sufficiently large). At this, the influence of clouds can be determined with the help of the developed method for the transmission spectrum interpretation, and the liquid-water content of clouds can be evaluated with an accuracy up to 15%. However, the accuracy of satellite measurements of general water content is about 8%.

7. The Use of Active Radar Aids

The possibilities of using active radar aids, both surface and airborne ones, for the purposes of geophysics are well enough known. Since in the near future the development of one multi-purpose radar which would allow to simultaneously obtain the whole complex of geophysical information answering the user's requirements is hardly possible, it seems worthwhile to make two types of satellite radars. The radars of the first type (radar-I) are used for the detection of clouds, precipitation and meteorological disturbance characteristic. They have an antenna with a scanning, pencil-beam pattern which allows a satisfactory selection of clouds and precipitation according to the distance, since the influence of the disturbing reflections from the surface essentially decreases. Moreover, such an antenna pattern makes it possible to achieve sufficient energy and geometric resolutions. The radars of the second type (radar-II) solve the remaining problems and have antennae with two flat-sector patterns which allow lateral observation of the earth's surface.

The radar energy resolution creating the possibility of non-homogeneity detection is determined by the radiated power P_t and the receiver sensitivity P_{min}. The analysis of the basic radar equation shows that the required values of P_t and P_{min} at present correspond to those technically achieved in the microwave range, which results in the realization of the detection of local non-homogeneities on the earth's surface.

The selection of the optimal wavelength range for radars-I shows that they should work on $\lambda=2$–3 cm [73, 32]. The results of the quantitative evaluation of the possibility of cloud and *precipitation* detection from satellites with a probability not less than 0.9 for the radar with $\lambda=3$ cm, sounding pulse length of 1 μsec, pattern width in the vertical and horizontal planes of 0.51° and 3.5°, attenuation coefficient of 0.5, $P_t=300$ kW, $P_{min}=10^{-14}$ for various values of the distance r, overlay coefficient K_3 and radar reflectivity of clouds $Z=\sum_i ND^6$ (where N is the number of particles in m^3, and D is their diameter) are shown in the Table I.

TABLE I

r (km)	200	400	600	800
K_3	0.5	0.3	0.1	0.08
Z (mm^6/m^3)	64	415	2880	5100
I (mm/hour)	0.46	15	5.08	7.25

The table also gives the values of precipitation intensity, the corresponding correlations $Z=200I^{1.6}$. It can be seen from the table that from low orbits almost all the precipitation can be detected, and above all the intensive precipitation connected with hurricanes and typhoons. Satellite determination of the quantitative characteristics of clouds and precipitation is a more difficult though feasible task.

Radars-II may also be used for the determination of the characteristics of *meteorological disturbance* [83, 52]. It is known that at the reception of the signals reflected from the sea's surface both a constant (mean) and a fluctuating component are observed which are called a coherent and a non-coherent component, respectively. The reflection coefficient $K(\beta)$ of the non-coherent component means the ratio of the power reflected in the given direction and the power which would be obtained in the case of isotropic reflection. Papers [83, 52] give the corresponding expressions for the relative dependence of $K(\beta)$ from the sighting angle and other factors. From the analysis of these expressions and experimental data obtained on $\lambda=3$ cm the difference in the effective cross section $\sigma(\beta)$ or $K(\beta)$ between slight and heavy disturbance can well be seen which allows one to qualitatively solve the inverse problem. Moreover, the quantitative analysis of $K(\beta)$ can furnish data on the mean (effective) steepness of waves and on the presence of small non-homogeneities upon the sea surface. The latter will make it possible to distinguish wind waves from swell, the wave crests of which are not broken and consequently produce no spray or foam. This is done from the value of the angle β at which $K(\beta)=0.5$. Then from the $K(\beta_m)$ value, where $\beta_m=40$–$50°$, the presence of small formations is inferred. The direction of wind near the sea's surface can also be determined: for this purpose the scanning of the antenna should take place not in the vertical plane, but in the plane inclined towards it at the angle of 10–15°. The maximum signal arrives in the direction opposite to that of wind.

The coherent component of the echo is observed only in the direction of the specular reflection, i.e. in our case with $\beta=0$. The amplitude of this component allows one to determine the wave height from its measured absolute value. Since this is difficult, it is reasonable to use the ratio of the coherent component to the non-coherent one. The practical realization of the latter method requires a second channel in the radar-I operating in the range of several meters [83].

Radars-II can be used for the obtaining of data upon the earth's surface characteristics. Since they should not depend upon weather conditions, the optimal range of their wavelengths is in the limits of 3–75 cm [17]. To increase the amount and quality of information, it is expedient to use multi-frequency and multi-polarizational radar systems allowing one to measure reflection coefficients with various sighting angles at one or several frequencies [17, 51, 61]. It can be expected that the statistical processing of the complex of such information will ensure with a high probability the identification of the majority of radar targets and the assessment of their characteristics.

The realization of a radar-II with the synthesized antenna pattern may result in the achievement of the geometric resolution of the order of 15 m [17]. Such a resolution is quite satisfactory for the solution of many geophysical problems, particularly for the radar *ice reconnaissance* [42, 50, 51]. It includes the detection of icebergs and the tracking of ice edges, the determination of ice compactness and form, as well as of the size of numerous leads and openings. Using the same method as that of determining disturbances, it is apparently possible to determine ice roughness. It should be noted that of ice fields it is easier to judge from the radar pictures than from photographs about the genetic type of ice, the structure of its surface, as well as about rifts covered with snow.

Another field for using radars-II is *hydrology*, particularly its basic part – that of surface water outlets. For this purpose large-scale basic surface charts (of the order of 1:25 000) are necessary, especially those of mountainous and other little investigated regions. Such a scale is ensured by modern airborne radars through lateral observation [51, 57], with the help of which it is possible to detect all the water river beds of the first order, almost all the tributaries of the second order and only large tributaries of the third order. This has made it possible to ascertain the correlations between topographic and radar data: the water-collection area, the parameter of the basin, the length of tributaries, the steepness of slopes, which correlations sometimes appear rather high, of more than 0.9 [51]. The clear distinction between the radar pictures of water surfaces and those of land, together with the radar pictures of precipitation zones will give the most valuable information for the purposes of high water and flood forecasts.

Aircraft radar pictures are widely used for geological purposes, particularly for the topographic survey, for the investigation of the unique features of the geological structure, and for the compilation of geological maps. It should be noted that for the mapping of unexplored or little-investigated territories the radar method is highly economical, and that in a number of cases it allows one to discern those features of the surface and basic rocks which are not revealed by photographs. The latter advantage is mainly due to the deep penetration of micro-radiowaves into the thickness of the upper layers of the earth, as well as to the fact that vegetation and snow cover scarcely affect the ability of distinguishing the underlying rocks [57, 15, 48]. It is stated that the method becomes more effective when a two-polarization radar operating in the mode of cross-polarization reception is used in combination with the infrared aerial photographic survey of the terrain.

Agriculture is another field of application of radars. This method of obtaining the required agricultural information should be particularly independent of weather conditions. Aircraft methods of obtaining agricultural information are based upon the use of a two-polarization radar, with the help of which field boundaries, areas and kinds of crops are determined. In particular, it has appeared possible to discern with a high probability the plantings of sugar-beet and the unsown regions of land [51, 53]. Further investigations are supposed to make it possible to determine the real sowing areas, the state of farming cultures, as well as to find out the reasons for the decrease of crop capacity, and the like. In this connection and for the characterization of the landscape as well, methods are of interest for the processing and presentation of the data of the two-polarization radar in the form of three-colour maps of vegetation (forests, meadows, swamps, etc.) with the help of the IDECS electronic system [57, 53].

The use of radio altimeters mounted on satellites enables one to determine the small-scale peculiarities in the geoid configuration and to affirm its large-scale features. Calibration difficulties arising in the process can be overcome by the determination of the height over seas and oceans. The accuracy of height measurements of 30 m achieved under the 'Saturn' program promises a successful solution of the above said problems [57].

The necessary frequency of obtaining information on precipitation, disturbance, etc. from low orbits can be achieved by the use of numerous satellites put into circum-polar circular orbits. Meanwhile, this problem can in principle be solved with the help of two or three satellites put into synchronous orbits of about 36 000 km high. Preliminary calculations show that radars operating on $\lambda = 2$ cm with the antenna aperture area of 10 m^2 give an elementary platform on the earth's surface of 70 km^2. To distinguish genuine signals caused by falling rain drops from the parasitic ones reflected by the earth's surface, a simple Doppler radar can be used [57, 30].

A combined active-passive radar system is particularly promising. By obtaining simultaneously reflection and emission characteristics of an object with the help of such a system, it is possible to nearly achieve optimum results in the identification and determination of the object's quantitative characteristics. Since the values of the reflection coefficient and the radio brightness temperature are measured, such a system operating on the proper wavelength enables already at present to obtain the temperature values for the seas' and oceans' surfaces corresponding to those of formula (1), as well as to considerably increase the probability of precipitation detection and the accuracy of their intensity determination, to notably improve the quality of geological, ice and other information.

The application of both simple and more complicated space radar instruments requires simultaneous development of special systems for the processing and presentation of the obtained data. Paper [57] contains a discussion of the aspects of such a system.

8. Conclusion

In conclusion, the importance should be noted of the problem of complex systems and satellite sensors for various forms of registering the electromagnetic field of the earth in different wavelength bands. This problem, because of its technical difficulties, is well suited to an orbital research laboratory operated, controlled and adjusted by human operators. The obtaining of complex data, mutual control and interaction of different sensors make every separate source of information more valuable and considerably increase reliability and the whole amount of information on the natural resources of the earth.

References

[1] Arking, A.: 1964, 'Latitudinal distribution of cloud cover from TIROS III photographs', *Science* **143**, No. 3606.

[2] Avaste, O. A.: 1963, 'The spectral albedo of the main types of the underlying surfaces in the near infrared spectral region', *Investigations in Atmospheric Physics*, No. 4, Tartu.

[3] Anonymus: 1966, *Aviation Week and Space Technology* **84**, No. 16.

[4] Badgley, P. C. and Vest, W. L.: 1966, 'Orbital remote sensing and natural resources', *Photogramm. Eng.* **32**, No. 5.

[5] Bristor, C. L., Callicott, W. M., and Bradford, R. E.: 1966, Operational processing of satellite cloud pictures by computers', *Month. Weath. Rev.* **94**, No. 8.

[6] Burns, E. A. and Lyon, R. T. P.: 1965, 'Instrumentation for a mineralogical satellite', in *Proc. XIV IAF Congress*, Paris, 1963, vol. 2, Warsaw.

[7] Carneggie, D. M. and Lauer, D. T.: 1966, 'Uses of multiband remote sensing in forest and range management', *Photogrammetria* **21**, No. 4.
[8] Chopra, K. P. and Hubert, L. F.: 1965, 'Karman vortex streets in wakes of island', *AIAA J.* **3**, No. 10.
[9] Clapp, P. F.: 1964, 'Global cloud cover for seasons using TIROS nephanalyses', *Month. Weath. Rev.* **92**, No. 11.
[10] Colwell, R. N.: 1966, 'Potential uses of a survey satellite in vegetation and terrain analysis', in *Proc. Am. Astronaut. Soc. on Practical Space Applications*, San Diego.
[11] Conover, J. H.: 1964, 'The identification and significance of orographically induced clouds observed by TIROS satellites', *J. Appl. Met.* **3**, No. 3.
[12] Conover, J. H.: 1963, 'Interpretation of clouds and mesoscale cloud patterns as seen from satellite altitudes', in *Proc. First Intern. Symp. on Rocket and Satellite Meteorology*, Washington, 1962, Amsterdam.
[13] Conover, J. H.: 1965, 'Note on the flora and snow cover distributions affecting the appearance of northeastern United States by TIROS satellites', *Month. Weath. Rev.* **93**, No. 10.
[14] Croning, J. F. and Clifford, C. D.: 1967, 'Interferometric determination of the spectra infrared emissivities of terrestrial surfaces', in *Proc. Symp. on Electromagnetic Sensing of the Earth from Satellites*, Polytechn. Inst. of Brooklyn, N.Y., 1965.
[15] Dellwig, L. F. and Moore, R. K.: 1966, 'The geological value of simultaneously produced like- and cross-polarized radar imagery', *J. Geophys. Res.* **71**, No. 14.
[16] Dulevich, V. Ye., Burenin, N. I., Melnik, Yu. A., and Korostylev, A. A.: 1964, *The theory of radar*, Sov. Radio, Moscow.
[17] Ellermeiter, R. D. and Simonett, D. S.: 1967, 'Imaging radars on spacecraft: a tool for studying earth, moon and the planets', in *Proc. Symp. on Electromagnetic Sensing of the Earth from Satellites*, Polytechnic Inst. Brooklyn, N.Y., 1965.
[18] Feigelson, Ye. M.: 1964, *Radiation processes in stratified clouds*, Nauka, Moscow.
[19] Fraser, R. S.: 1964, 'Apparent contrast of object on the earth's surface as seen from above earth's atmosphere', *J. Opt. Soc. Am.* **54**, No. 3.
[20] Fritz, S.: 1963, 'The variable appearance of the earth from satellites', *Month. Weath. Rev.* **91**, No. 10–12.
[21] Fritz, S: 1963, 'Snow surveys from satellite pictures', in *Proc. First Intern. Symp. on Rocket and Satellite Meteorology*, Washington, 1962, Amsterdam.
[22] Fritz, S.: 1964, 'Pictures from meteorological satellites and their interpretation', *Space Sci. Rev.* **3**, No. 4.
[23] Fritz, S.: 1965, 'The significance of mountain lee waves as seen from satellite pictures', *J. Appl. Met.* **4**, No. 1.
[24] Garrison, W. L., Dacey, M. F., Dacey, D. F., Marble, D. F., Alexander, R., and Bailey, W.: 1966, 'Data system requirements for geographic researches', in *Proc. Third Goddard Memorial Symp. on Scientific Experiments for Manned Orbital Flight*, Washington, 1965, *Sci. Techn. Ser.* **4**.
[25] Gates, D. M., Keegan, H. J., and Weidner, V. R.: 1966, 'Spectral reflectance and planetary reconnaissance', in *Proc. Third Goddard Memorial Symp. on Scientific Experiments for Manned Orbital Flight*, Washington, 1965, *Sci. Techn. Ser.* **4**.
[26] Haffner, W.: 1966, 'Vier kleinmassstäbige Luftbilder aufgenommen vom Wettersatelliten Nimbus A', *Erdkunde* B.20, H.2.
[27] Holliday, C. T.: 1954, 'The earth as seen from outside atmosphere', in *The Solar System*, vol. 2, *The Earth as Planet*, Chicago.
[28] Hope, T. R.: 1966, 'Path of heavy rainfall photographed from space', *Bull. Am. Met. Soc.* **47**, No. 5.
[29] Hyatt, H. A.: 1967, 'Airborne measurement of microwave emission from the earth's surface and atmosphere', in *Proc. Symp. on Electromagnetic Sensing of the Earth from Satellites*, Brooklyn, N.Y., 1965.
[30] Katzenstein, H. and Sullivan, H. A.: 1963, 'A new principle for satellite-borne meteorological radar', in *Proc. Tenth Weather Radar Conference*.
[31] Kennedy, J. M. and Sakamoto, R. T.: 1967, 'Passive microwave measurements of snow', in *Proc. Symp. on Electromagnetic Sensing of the Earth from Satellites*, Brooklyn, N.Y., 1965.
[32] Kigler, J. E. and Krawitz, L.: 1960, 'Weather radar observation from earth satellites', *J. Geophys. Res.* **65**, No. 9.

[33] Kock, W. P. de: 1966, 'Geologising from space', *S. Afr. Mining Eng. J.* **77**, No. 3824.
[34] Kondratyev, K. Ya.: 1963, *Meteorological satellites,* Gidrometeoizdat, Leningrad.
[35] Kondratyev, K. Ya.: 1967, 'Meteorological investigations from manned space vehicles', in *Proc. XVII IAF Congress*, Warsaw.
[36] Kondratyev, K. Ya., Borissenkov, Ye. P., and Morozkin, A. A.: 1966, 'Practical use of meteorological satellite data', Gidrometeoizdat.
[37] Kondratyev, K. Ya., Gayevsky, V. L., Gusseva, L. N., Ter-Markaryantz, N. E., and Zhvalev, V. F.: 1968, 'Meteorological interpretation of infrared pictures of the earth from space', *Pure Appl. Geophys.* 1968.
[38] Kondratyev, K. Ya., Mironova, Z. F., and Otto, A. N.: 1965, 'The spectral albedo of the natural underlying surfaces', *Problems of Atmospheric Physics*, No. 3, Proc. Leningrad State Univ., Leningrad.
[39] Kondratyev, K. Ya. and Smokty, O. I.: 1968, 'Geophysical investigations on manned spacecrafts', *Pure Appl. Geophys.* **54**.
[40] Krinov, Ye. L.: 1947, 'Spectral reflectivity of natural formations', *Proc. Acad. Sci. U.S.S.R.*, Moscow-Leningrad.
[41] Lenior, W. B.: 1967, 'Remote sounding of the atmospheric temperature by microwave measurements', in *Proc. Symp. on Electromagnetic Sensing of the Earth from Satellites*', Brooklyn, N.Y., 1965.
[42] Loshchilov, V. S. and Shkolnikov, V. I.: 1965, 'The experience of the application and the prospects of the radar method in ice reconnaissance', in *Application of Radiophysics Methods in Oceanographic and Ice Investigations*, Leningrad.
[43] Low, D. S. and Braithwaite, J. G. N.: 1966, 'A spectrum matching technique for enhancing image contrast', *Appl. Opt.* **5**, No. 6.
[44] Low, D. S. and Shay, J. R.: 1967, 'Agricultural sensing from orbiting satellites', *Proc. Symp. on Electromagnetic Sensing of the Earth from Satellites*, Brooklyn, N.Y., 1965.
[45] Lowe, J. D. and Coleman, I.: 1967, 'The application of multiplescan interferometry to the measurement of infrared spectra of minerals', in *Proc. Symp. on Electromagnetic Sensing of the Earth from Satellites*, Brooklyn, N.Y., 1965.
[46] Lowman, P. D.: 1965, 'Photography of the Earth from sounding rockets and satellites', *Photogramm. Eng.* **31**, No. 1.
[47] Lowman, P. D.: 1966, 'Photography from space-geological applications', *Ann. N.Y. Acad. Sci.* **140**, No. 1.
[48] MacDonald, H. C., Brennan, P. A., and Dellwig, L. E.: 1967, 'Geologic evaluation by radar of NASA sedimentary test site', *IEEE Trans. Geoscient. Electron.* vol. GE-5, No. 3.
[49] Mitschem, Th. W.: 1959, 'Viking rocket photograph of Arizona, California and northern Mexico west of Tucson, Arizona', in *Geol. Soc. Guidebook*, vol. 2.
[50] Mizinov, P. I. and Tsurikov, V. L.: 1965, 'The generalization of the experience of the application of radars to oceanographic observations', in *Application of Radiophysics Methods in Oceanographic and Ice Investigations.* Leningrad.
[51] Moore, R. K., and Simonett, D. S.: 1967, 'Potential Research and Earth Resource Studies with Orbiting Radars: Results of Recent Studies', AIAA Paper, No. 67-767, Anaheim, California.
[52] Moore, R. K.: 1967, 'Measuring sea state and estimating surface winds from Polar Orbiting Satellite', in *Proc. Symp. on Electromagnetic Sensing of the Earth from Satellites*, Brooklyn, N.Y., 1967.
[53] Morian, S. A. and Simonett, D. S.: 1967, 'K-band radar in vegetation mapping', *Photogramm. Eng.* **33**, No. 7.
[54] Morian, S. A. and Simonett, D. S.: 1966, 'Vegetation with radar imagery', in *Proc. Fourth Symp. on remote Sensing of Environment*, Chicago, 1965.
[55] Morrison, A. and Bird, J. B.: 1965, 'Photography of the Earth from space and its non-meteorological applications', in *Proc. Third Symp. on Remote Sensing Environment,* Michigan, 1964.
[56] Morrison, A. and Chown, M.: 1965, 'Photographs of the Western Sahara from Mercury MA-4 Satellite', *Photogramm. Eng.* **32**, No. 4.
[57] Myaskovsky, O. M., Sadov, A. V., and Fedin, V. T.: 1966, 'The possibilities of using thermal and radar aerial photography in geology', Survey, Inst. Experimental Meteorol. Synoptics, Moscow.
[58] Nikolayev, A. G. and Pertsev, S. V.: 1964, 'Thermal radar', Sov. Radio, Moscow.

[59] 'Oceanography from space', in *Proc. Conf. on the Feasibility of Conducting Oceanographic Explorations from Aircraft, Manned Orbital and Lunar Laboratories*, Massachusetts, 1965.
[60] 1956, 'Picture of the month', *Month. Weath. Rev.* **94**, No. 6.
[61] Pierson, W. J., Scheps, J. B. and Simonett, D. S.: 1966, 'Some applications of radar return data to the study of terrestrial and oceanic phenomena', in *Proc. Third Goddard Memorial Symposium on Scientific Experiments for Manned Orbital Flight*, Washington, *Sci. Techn. Ser.* **4**.
[62] Popham, R. W. and Samuelson, R. E.: 1965, 'Polar exploration with Nimbus meteorological satellite', *Arctic* **18**, No. 4.
[63] Rall, L. L.: 1966, 'Geographic data processing', *Photogramm. Eng.* **32**, No. 6.
[64] 1966, Report of the Panel on International Meteorological Cooperation, Washington.
[65] Rosenfeld, A., Fried, Ch., and Orton, N.: 1965, 'Automatic cloud interpretation', *Photogramm. Eng.* **31**, No. 6.
[66] Schneider, S.: 1965, 'Die Satellitenphotographie und ihre Anwendung für die geographische Forschung', *Bildmessung und Luftbildwesen* **33**, J., H.2.
[67] Singer, S. F.: 1962, 'Forest fire detection from satellites', *J. Forestry* **60**, No. 12.
[68] Singer, S. F. and Popham, R. W.: 1963, 'Non-meteorological observations from weather satellites', *Astronaut. Aerospace Eng.* **1**, No. 3.
[69] Singer, S. F. and Williams, G. F.: 1968, 'Microwave detection of precipitation over the surface of the ocean', *J. Geophys. Res.* **73**, No. 10.
[70] Sonechkin, D. M.: 1962, 'On the deciphering of TV cloud pictures obtained from satellites', *Meteorol. Hydrol.*, No. 9.
[71] Sonechkin, D. M. and Tryapitsina, I. S.: 1967, 'Some results in the investigation in the cloudiness statistical structure', in *Trans. Hydrometeorol. Centre U.S.S.R.*, II, *Problems of Satellite Meteorology*.
[72] Stealin, D. H.: 1967, 'Measurement and interpretation of the microwave spectrum of terrestrial atmosphere', in *Proc. Symp. on the Electromagnetic Sensing of the Earth from Satellites*, Brooklyn, N.Y., 1965.
[73] Stepenenko, V. D.: 1966, 'Radar in Meteorology', Gidrometeoizdat, Leningrad.
[74] Stogryn, A.: 1967, 'The apparent temperature of the sea at microwave frequencies', *I.E.E.E. Transactions on Antennas and Propagation*, Vol. AP-15, No. 2.
[75] Taggart, C. I.: 1963, 'Ice reconnaissance from space satellites', *Canad. Aeronaut. Space J.* **9**, No. 3.
[76] Chapursky, L. I.: 1966, 'Experimental investigations in the spectral brightness characteristics of clouds, atmosphere, and underlying surface in the wavelength interval of 0.3–2.5μ', *Trans. Main Geophys. Observatory*, No. 196.
[77] Troitsky, V. S., Zhevakin, S. A., and Tseitlin, N. M.: 1958, 'Atmospheric radioemission and the investigation in the absorption of centimetric waves', *Bull. Higher Institutions U.S.S.R., Radiophysics* **1**, No. 2.
[78] Vinogradov, B. V.: 1966, 'Space photography in the geographic investigation of the earth', *Bull. Geograph. Soc. U.S.S.R.* **98**, No. 2.
[79] Vinogradov, B. V.: 1967, 'The landscape regionalization by means of air and space photography'. Papers presented to Meeting of the Commission on Interpretation of Aerial Photographs of I.G.U., Ottawa, 1967, Zürich.
[80] Wark, D. Q. and Popham, R. W.: 1963, 'The development of satellite ice observation technique', *Proc. First International Symp. on Rocket and Satellite Meteorology*, Washington, 1962, Amsterdam.
[81] Wilkerson, J. C.: 1967, 'Some Nimbus-2 satellite observations of the Gulf Stream', *Trans. Am. Geophys. Union* **48**, No. 1.
[82] Wilson, R. C.: 1967, 'Space photography for forestry', *Photogramm. Eng.* **33**, No. 5.
[83] Zubkovich, S. G. and Marchenko Yu. I.: 1965, 'The possibilities of the determination of some characteristics of sea disturbance by means of radiotechnical instruments mounted on vehicles', in *Application of Radiophysics Methods in Oceanographic and Ice Investigations*, Leningrad.

SPACECRAFT FOR OCEAN EXPLORATION

EDWARD WENK, JR.

National Council on Marine Resources and Engineering Development, Washington, D.C., U.S.A.

Abstract. Spacecraft oceanography is in its infancy, but evidence to date suggests that it has great potential for ocean exploration. The synoptic scanning from orbiting platforms can perform the work of many ocean surface sensors in observing the large-scale oceanic processes. Satellite scans offer unique possibilities for broad, even global, unified synoptic analyses. Such observations are ideally suited to the upcoming International Decade of Ocean Exploration for the 1970's. The Decade is envisioned as a concerted, long-term, international cooperative program of ocean exploration on a world-wide basis.

Although oceanic data obtained from space today is less quantitative than surface observations, the coming decade will undoubtly see improvements. Combined with data obtained from ships and buoys, spacecraft observations will provide a new and useful capability for meeting the challenge of the oceans.

1. Spacecraft for an International Decade of Ocean Exploration

From the earliest days of scientific oceanography, it was recognized that observations taken only from ships were limited. The vastness of the oceans and their variation with time suggested that large scale oceanographic processes could not be satisfactorily observed and understood from scattered measurements. Information about large-scale oceanographic processes depends upon the analysis of data obtained from observations taken in a variety of locations and times.

Spacecraft or aircraft platforms afford new and superior observation points. Photographic and spectral scanning from such platforms can provide broad, even global coverage. For some measurements, these scans would be equivalent to those from thousands of individual ocean surface sensors. While data acquired from space today are less precise compared with corresponding surface measurements, a sensing system with global coverage, reporting as often as desired and with increasing sophistication and accuracy, certainly affords a powerful new tool for synoptic and global analyses.

2. Potential Uses of Spacecraft Observations

The recognized need for broad ocean observations and developing technology converged about 10 years ago, and spacecraft oceanography was born. Its progress has been slow. Early last year the Marine Sciences Council considered the current state of the development and its promise for new uses in the oceans. Initial studies indicated the following potential spacecraft observations would be made from space:

ocean surface and, indirectly, subsurface temperatures and currents;
sea state;
indices of biological activity;
coastal processes;

S. F. Singer (ed.), Manned Laboratories in Space.

hydrography and bathymetry;
sea ice;
air/sea interactions;
mean sea level;
salinity and water density near the surface and in the subsurface.

Examples illustrate these uses.

Temperature measurements. Oceanographers have long studied the thermal structure of the oceans, for it is fundamental to all marine processes including the migration of marine life. Large-scale thermal features of the ocean are dynamic and ordinarily can be detected and studied only with repetitive measurements of wide areas. Sea surface temperature variations, even very subtle ones, have been detected from aircraft in the infrared and microwave regions of the spectrum. Under cloud-free conditions, it has been possible to map and measure areas of strong thermal contrast. Nimbus infrared radiometer images have proved useful, for example, in tracing the Gulf Stream, mapping sea surface temperature gradients and the positions of current boundaries, meanders, and eddies in the North Atlantic Ocean.

Coupled with that success we have learned through aircraft photos that fishery stocks ordinarily do not cross the thermal interface of the Gulf Stream. As a consequence, when the Icelandic fishing fleet lost its herring stock in 1968 and was threatened with economic disaster, an infrared reconnaissance by United States Navy aircraft located the meandering edge of the Gulf Stream and made it possible for the fishermen to move to unexpected areas where they rediscovered their lost fishery. The Gulf Stream edge may also be identified by the accumulation of sargassum, visible in aircraft photos. Work has been initiated to relate sea surface temperatures obtained from radio metric data to those taken from onsite measurements.

Sea state. Understanding of the energy processes between the sea and the atmosphere has long been hampered by inadequate data on wave growth and decay. For the most part, in the absence of direct observations, oceanographic forecasters have had to infer sea state conditions from wind reports.

Several methods have now been advanced for detecting sea state from orbital altitudes. All are based on changes in reflective properties of the ocean surface stirred by varying wind and sea conditions. Sunlight scattered at the sea surface – sun glint – can be observed both from aircraft and satellites as a measure of sea state. Non-glitter portions of the surface have also been studied for sea state estimates. Color variations are often indicative. Color photography by aircraft, as recorded in the Apollo 501 and Apollo 502 and Gemini flights, show a high correlation between nearshore ocean depth contours and film density gradients. For example, a lighter shade of blue in the sea may result from high sun reflectance related to high sea state; similarly darker water may suggest low sea state.

Still more sophisticated optical techniques have been applied. But they are often limited by the cloud cover which veils many parts of the ocean. Passive microwave radiation and radar reflection are now being considered for such observations. The Navy has used neon-helium lasers to profile ocean surface waves.

Indices of biological activity. The efficiency of fishing operations would sharply increase if the search time for productive fishing grounds were reduced. Forecasting such locations requires some prior knowledge of environmental conditions and the abundance of fish. It is known that temperature regimes, especially, provide such valuable indications. These regimes may be observed from space. There is accumulating evidence also that under favorable conditions large fish schools may be directly detected from space. Aircraft-borne sensors have detected the bioluminescence of living organisms, and the technique may prove to be one of the most useful means of orbital fish detection.

Sea ice. The International Ice Patrol regularly uses satellite ice data from various polar orbiting meteorological satellites to augment information acquired by ships and aircraft. Infrared and microwave imagery has revealed the texture of ice filled waters; even individual icebergs have now been observed. Recently efforts have been made to improve computer processing of sea ice imagery and to develop infrared techniques for ice mapping. The Navy has used ruby lasers on aircraft for imaging sea ice topographic profiles.

Coastal processes. The processes of the coast involve the complex phenomena where the land and sea meet or where fresh water and salt water meet. The Coastal Zone is also where the people and the oceans meet, and thus where an intense amount of economically significant activity is based. Space photographs taken during the Gemini flight series indicate the potential use of coastal processes observations – to reveal effluent patterns, currents and bottom topography. Sediment patterns and coastal depths can be measured through density contrast color printing simulating multi-spectral photography. It is also possible to observe water depth penetration.

Hydrography and bathymetry. Spacecraft are potentially useful in mapping. High altitude photographs superimposed on navigational charts sometimes reveal discrepancies. Where surface data are available, the charts are usually very accurate. But where such data are missing, the charts often show errors which can only be corrected by employing a mosaic of over some 100 aerial photographs.

For example, ocean photographs from space showing different shades of blue reveal different depths. On the theory that the blue shade is proportional to water density, depth contours can be plotted through use of a density separation process. By superimposing photographs, it is possible to construct an isodensity photograph which in turn corresponds to a single contour of depth for charting purposes. Little known subsurface features as shoals and reefs, currents and sedimentation dispersions are also discernible.

Navigation. Satellites have also been valuable in the well developed field of navigation and ocean communications. Last year the United States Navy released for civilian use the Transit Satellite System which improves navigational accuracies for ships to less than one-tenth of a mile. Also, the Omega Position Location Experiment (OPLE) improves many-fold the positioning accuracies of ships at sea. Both systems will facilitate ocean exploration. The 'Geos' series of satellites are potentially valuable tools for global ocean level measurements.

3. Communication Satellites

As to communication satellites, it is expected that these systems will eventually be employed to interrogate large numbers of surface sensors located on both ships and buoys–simultaneously fixing their positions, recording data transmissions twice a day, and relaying them to ground terminals. A polar orbiting Nimbus flight test for such a system is scheduled for 1969.

To link the capabilities from space and the needs for ocean data does not happen spontaneously. Leadership, funds and institutional arrangements are required to accomplish the desired marriage. Toward this objective,

The National Aeronautics and Space Administration is coordinating federal work in sensor technology, techniques for space oceanography, and testing new developments; NASA will use the Polar Nimbus satellite for ocean temperature sensing;

The Environmental Science Services Administration (ESSA) of the Department of Commerce is to employ an advanced camera system and high resolution scanning radiometer, pickaback, in the TIROS M and advanced TIROS operational satellites for mapping sea and lake ice and surface temperatures, locating major currents, and estimating the state of the sea; and

The Bureau of Commercial Fisheries (BCF) of the Department of the Interior is to expand studies of the usefulness of satellite radiometric data and photographs in locating high density fisheries.

The various polar-orbiting meteorological satellites, TIROS, Nimbus, and Gemini, have shown remarkable ocean features, in considerable detail, including subsurface elements, air-sea interactions such as hurricanes, evidence of fish schools and other biota, and effluents.

Stationary satellites, ATS, with much reduced resolution, can scan an entire ocean at once to determine grosser features and cloud cover, often indicating special ocean behavior. Manned spacecraft such as Apollo will be especially beneficial for they will permit optical photography in the visible and near infrared regions of high resolution.

In the expectation that the Apollo flights will make possible opportunities for photography of marine areas, regions of interest are being considered. Among them are the following:

The Caribbean Sea. The United States has supported the Dutch proposal that the Intergovernmental Oceanographic Commission of UNESCO sponsor an international collaborative investigation of the Caribbean and adjacent regions beginning in 1970. Apollo photography could reveal oceanographic features of interest to many scientists and could establish the baseline from which to launch the cooperative venture. Also, it could reduce some of the uncertainties about the precise location of some of the smaller Caribbean Islands. Spacecraft played a role in the air-sea interaction research of the Barbados Oceanographic Meteorological Experiment (BOMEX) which was conducted from May to July 1969 east of the Barbados.

The Chesapeake Bay and its major tributaries. The Marine Sciences Council in 1968 initiated a major United States Government interagency case study of the Chesapeake Bay to identify the potential role of science and technology in promoting the optimal use of an estuary. In the study the Army Corps of Engineers is promoting a three-dimensional hydraulic model of the Bay on a horizontal scale of 1:1 000, to aid development of a detailed framework to predict effects of various measures on the Bay. Space photography of the Bay would confirm and supplement fragmentary topographic information available from a wide variety of maps, ground photography, and other sources which are being used as the basis for designing the model. It would also provide a valuable point of reference in analysing future changes of the Bay.

Trust Territories. Maps and charts of the United States Trust Territories, such as the Marianas, and Marshall Islands, are deficient in many respects. Space photography or video pictures could assist in delineating coastal areas and upgrading maps and charts; it would thus assist maritime, fishery, and other interests.

Eastern end of the Mediterranean Sea. The construction of the Aswan Dam has resulted in profound changes in the physical and biological character of the Eastern Mediterranean Sea. These changes are continuing. They have a direct impact on the fishery resources available to both Israel and the Arab countries. Satellite photography would be a useful aid to many nations in understanding the phenomena of silt deposition, changing circulatory patterns, and sea features in the Eastern Mediterranean.

Northern and Northwest Edge of the Arabian Sea. This region is believed to be one of the richest untapped fishing areas of the world. Yet little is known about many of its oceanic features. Gemini photographs have already demonstrated that satellite photography may be directly applied to analyzing fish potentials. Additional photography of this area would provide further opportunities to advance the state-of-the-art while providing more precise topographical and other physical data.

4. The International Decade of Ocean Exploration

Advanced spacecraft technology can serve as a vital tool for future exploration of the oceans. But if it is to receive maximum use, it is necessary, as in all uses of our advancing ocean technology, to relate it to society's needs. Society's priority concerns must be identified; then new ideas, capital and enterpreneurship must be mobilized to use the seas to meet these needs. It is clear that before man can use the seas, he must explore it to understand its vagaries and potential. The sea is so vast and the exploratory task so great that only through use of our advanced technology and participation by many nations will we acquire the needed knowledge in a reasonably short interval.

Recognizing this, former President Johnson proposed on March 8, 1968 the launching of "an historic and unprecedented adventure – an International Decade of Ocean Exploration for the 1970s". Activities during the Decade of Exploration could expand cooperative efforts among scientists from many nations to investigate the deep ocean and to increase knowledge of ocean resources to aid in meeting critical world food needs, bringing mankind closer to using new mineral and fossil fuel sources. The Decade is envisioned as a period of intensified collaborative planning among nations and expansion of national exploration capabilities, followed by implementation of national and international programs of oceanic research and resource exploration so as to assemble a far more comprehensive knowledge of the sea in a reasonably short time.

Other nations were invited to share in developing the promise of the sea. The response of nations and international organizations has been highly favorable. In the fall of 1968, the United Nations General Assembly adopted resolutions supporting the proposal for an International Decade of Ocean Exploration within the framework of a comprehensive long-term program of scientific investigation of the oceans. It called on the Inter-governmental Oceanographic Commission of UNESCO to play a leading role in coordinating the program.

The development of a multinational framework for planning and cooperation is essential. But so too is the identification of specific projects to be carried out and individual national commitments to specific exploratory programs and cooperative endeavors. In the United States the Marine Science Council is responsible for developing the national plans and for coordinating their implementation. The industrial and scientific communities are being asked to join both in planning and in conducting the program. For this purpose the National Academies of Sciences and of Engineering have considered and reported on the scientific and engineering project needs they foresee for the Decade.

Spacecraft oceanography lends itself to a concerted, long-term, cooperative program of world-wide ocean exploration such as that proposed in the Decade. Flying platforms are effective observation stations for research and survey, to collect data on a broad regional and world ocean basis and to explore living resources and ocean processes.

Among the types of Decade projects in the living resource area, spacecraft might assess fishery stocks useful to man in uncharted regions of the world ocean, assess current use of fishing stocks, and seek knowledge related to the environment of living resources to permit greater efficiency in their capture and conservation.

As to ocean process studies, satellite observations can study the scales of motion in the sea and can obtain data about the dynamics of ocean current systems. They can investigate surface boundary processes such as the growth and propagation of ocean waves and study the evolutionary processes of ocean basins.

Spacecraft observations in a Decade program can be extremely helpful to interested less developed nations, in assisting them to map selected areas of their shorelines and continental shelves and in surveying coastal fishery resources.

Thus, such advanced technology magnifies the capability of man's senses many times and makes it more feasible for him to turn the resource potential of the sea to practical use. Spacecraft oceanography exemplifies this new technological readiness. To utilize it effectively we must draw on the expertise of our aerospace industry and scientists. This advanced technology is also available in our new research ships, deep ocean buoy technology, research submersibles, and modern data processing equipment. One of the tasks of the International Decade of Ocean Exploration will be to determine the optimum mix of these sensing platforms in exploring the oceans and deciding what tradeoffs can be made between ship surveys and observations from other platforms.

The United States has been able to develop unique spacecraft capabilities. It plans to make available to nations participating in the Decade further information in the form of atlases constructed from space data that could be superimposed on data collected at sea in projects mutually agreed upon internationally.

5. The Future

The promise of the future use of spacecraft oceanography is an auspicious one. It is hoped that space scientists in the U.S. and elsewhere will examine ocean exploration programs to see how spacebased observations will turn our expanded program of ocean investigation into an expedition of the 20th century.

Through use of spacecraft and other advanced devices today's little known marine environment may yield great benefits:

a growth in offshore industrial development of food, fuel, minerals;

a growth in world maritime trade;

a growth in ocean related sports and recreation;

a growth in our efforts to utilize our coastline and to maintain the natural heritage of its beauty;

a growth in use of the oceans as a scientific laboratory;

a growth in the importance of undersea capabilities in maintaining the peace.

Several years ago, the possibility of spacecraft oceanography seemed at best a dream. Yet today, while still in its infancy, the potential of this new means of investigation is real and encouraging. In the years ahead this new instrument will develop even more. Used in conjunction with surface ships, buoy networks and improved data management, it could open a new and more comprehensive view of the world oceans. The oceans, more than a geographical frontier, can prove to be a frontier in which we harness our technological capabilities in an effort to convert the ocean, as we have the land, to man's needs.

Bibliography

Ewing, Clifford C. (ed.): 1965, *Oceanography from Space*, Woods Hole Oceanographic Institution, Woods Hole, Mass.

Space Applications Summer Study, 1967, Volume I., Prepared by the National Academy of Sciences, National Research Council for the National Aeronautics and Space Administration.

United States Activities in Spacecraft Oceanography, The National Council on Marine Resources and Engineering Development, October 1, 1967, Superintendent of Documents, U.S. Government Printing Office, Washington, D.C.

Weaver, Kenneth F.: 1969, *National Geographic* **135**, 46–73.

Man's Geophysical Environment – Its Study from Space, a report to the Administrator of ESSA, prepared by an ESSA Task Force, chaired by J. P. Kuettner, U.S. Printing Office, Washington, D.C. 20402.

EARTH RESOURCE OBSERVATIONS FROM AN ORBITING SPACECRAFT

W. T. PECORA

Director, U.S. Geological Survey, U.S.A.

Abstract. Gemini photography and other satellite images have demonstrated some of the advantages of small-scale views of the earth from space. Long-lived satellites providing uniform repetitive coverage will add the dimension of time to our interpretive process and aid in better understanding the dynamics of earth.

Geologists, long hampered by their earth-bound approach of studying details and extrapolating to regional and world-wide scale, look forward to receiving images of the earth from space that will enable them to construct photo-image maps of the continents and the surrounding shelves. Such maps will enable them to focus inward on problems of interest and will assist in classifying terrain or recognizing anomalies that warrant further investigation. Not only will geologists be able to study the major mountain chains and their various structural elements in their regional, continental, or world-wide context, but they will in time be able to compare the surface changes that take place through the years. Volcanic eruptions, earthquakes, forest fires, and floods and their effects on erosion will be monitored and assessed from space in a rapid and timely manner. This should lead to improvements in prediction, assistance, and recovery techniques required in disaster areas.

Other time variant phenomena such as shore processes, estuarine changes, delta formation, sand dune development, and the growth or destruction of atolls will be permanently recorded and their changes mapped. Many of these geologic applications will provide information for further research and engineering planning related to land utilization, population centers, and urban growth.

1. Introduction

One of the basic problems facing the world today is the fact that basic mineral materials, land for urban and industrial development, and water resources, are being consumed at a rate that continues to rise as our world population acquires increasingly greater demands. These demands can be met in the future only if we can devise more efficient ways of finding new resources at a faster pace, and conserving those we already have.

Aerial photography and airborne geophysical surveys with geological interpretations have increased the rate at which the world's natural resources can be assessed or discovered, and have been widely used by geologists and geographers during the last 30 years. Recently, experimental color photographs of the earth's surface taken from short-term orbiting satellites have revealed the value of orbital photography in analyzing the earth's surface. Study of these photographs has suggested guidelines for the design of long-lived operational satellites, not only for photography, but for other remote sensor observational techniques.

2. Mercury-Gemini Photography

For geologists, hydrologists, geographers, and agriculturalists, involved in the evaluation of natural resources, the most significant achievement of the Mercury-Gemini

S. F. Singer (ed.), Manned Laboratories in Space.

series is the demonstration that scientifically useful photographs of the earth's surface can be taken from orbital altitude. This achievement is particularly significant in view of the fact that the primary objective of these missions was to evaluate the performance of man and a life support machine in space; therefore, the photography that was done was incidental to other tasks. For example, cameras were hand-held, resolution did not exceed 250 to 300 feet, and most of the photographs were taken as 'targets of opportunity' presented themselves and other duties permitted.

Despite these shortcomings, early analyses have shown several unique qualities of orbital photographs which are of direct benefit to geologists. These qualities include (1) the synoptic overview of a large area within a single exposure frame, (2) superior water penetration compared to conventional aerial photographs, and (3) near orthographic positioning of topographic and cultural features.

A. SYNOPTIC OVERVIEW OF A LARGE AREA

Largely on the basis of the work in tropical rain forest areas in South America, the U.S. Geological Survey (Hemphill, 1958) more than 10 years ago showed that small-scale aerial photographs (1:50 000-scale or smaller) are advantageous in some areas because they permit a continuity of observation not possible from photographs of larger scale. This work suggested that aerial photographs of even smaller scale than those generally available during the 1950's would be desirable for many geologic mapping problems.

Mercury and Gemini photographs acquired since 1960 clearly show that the regional significance of some geologic and physiographic features is more readily recognized and interpreted where viewed collectively on a single orbital photograph than on a conventional mosaic of aerial photographs of larger scale where thousands of prints would be required for the same area. Examples of features revealed by virtue of the synoptic overview provided by orbital photographs are shown in Figure 1.

Figure 1 is an orbital photograph covering about 7000 square miles in the Salt Range and Potwar Plateau, a semi-arid region of West Pakistan underlain by folded sedimentary rocks (Hemphill and Danilchik, 1968). Several fold axes can be seen in the northern part of the Potwar Plateau, and this region of relatively close folding adjacent to the Himalayan foothills contrasts markedly with the extensive homoclinal area in the southern part of the Potwar Plateau, farther from the region of maximum structural disturbance.

The north-trending fault zone north and east of the Indus River (Figure 1) also shows clearly on the orbital photograph, and is one of the major boundary faults south of the Himalayan foothills. Predominant movement is horizontal with the eastern block, comprising the Salt Range and Potwar Plateau, moving southward.

Close inspection of the central part of the photograph shows distinctive alignment of stream segments that are probably related to fractures in the underlying bedrock. Some of these alignment sets are several tens of miles in length but their existence was not previously known.

Fig. 1. Gemini photograph of the Salt Range and Potwar Plateau, West Pakistan, showing fold axes, major faults, and alignments of stream segments that are probably related to fractures in the bedrock. Arrows indicate areas possibly affected by 'water-logging'.

Light-toned areas in the southern part of the photograph may be evidence of 'water-logging', a process common to the Indus basin and other parts of the world where leakage from irrigation canals saturates adjacent areas; subsequent evaporation leaves behind a saline concentrate which renders previously arable land unfit for cultivation.

Delineation of major fold, fault, and fracture systems in a single integrated 'overview', as illustrated in Figure 1, is a new tool for geologists to use in achieving a better understanding of regional tectonic processes and the relation of individual structural features to their origin. Position and orientation of fracture systems as shown by alignments in topography and drainage pattern is information of direct importance in prospecting for commodities such as oil and water where deposits may be structurally controlled and hence discoverable.

B. WATER PENETRATION

Images of bottom features in some lakes and estuaries can be more readily achieved from photographs taken from a satellite than from an aircraft at lower altitude. Effective water penetration is greater from orbital altitude because, with the narrower angular field of view of most space cameras, the cone of rays is more nearly normal to the water surface, and the amount of light reflected and refracted at the air-water interface is thereby minimized.

Figure 2 is a Gemini photograph of Lake Chad at the southern edge of the Sahara Desert. Submerged sand dunes along the eastern shore exceed 3500 square miles. The western limit of the dunes probably approximates the diminished extent of the lake during a past dry period. General outline of the dunes have persisted despite wave action although infilling can be observed between the submerged dunes in the southern part of the photograph. The general orientation of the dunes is northwest, and is particularly apparent in the southern part of the lake near *A*. Distinct from this orientation is a northeast-trending lineation superimposed on the dunes in the lake at *B* and on-shore near *C* and *D*. The specific feature or features manifesting these

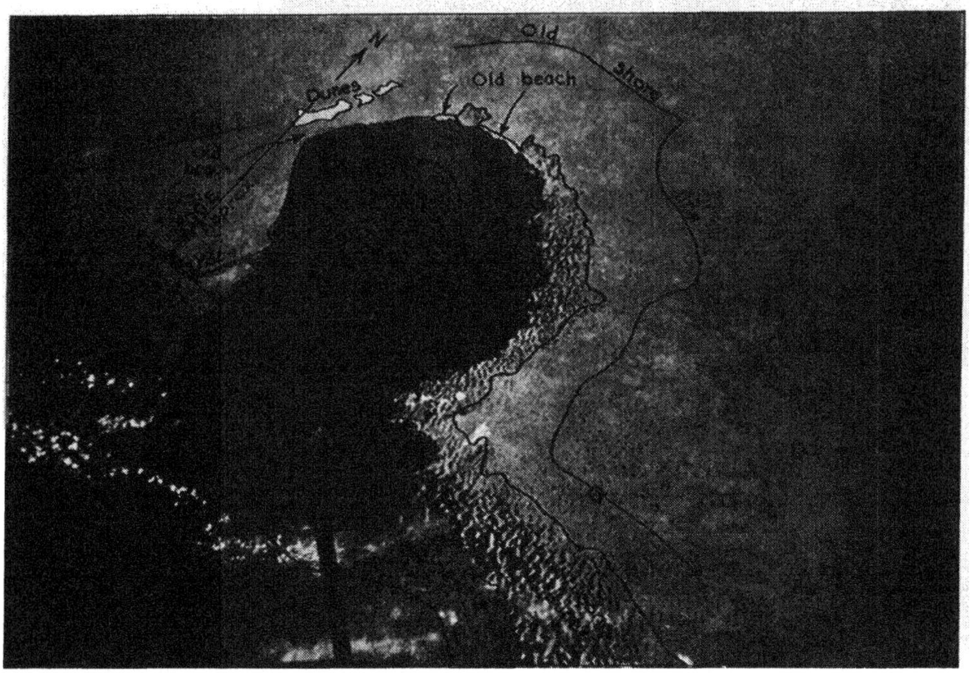

Fig. 2. Gemini photograph of Lake Chad, Republics of Chad, Niger, and Nigeria, showing submerged sand dunes and other physiographic features.

lineations is unclear; however, the lineations are probably related to the wind direction which prevailed when the dunes were deposited. The photograph not only illustrates the abundance of bottom detail that is discernible, but also shows how lineations in the lake bottom may be related to on-shore features more than 100 miles away.

C. ORTHOGRAPHIC QUALITIES

A series of 11 photographs acquired by Gemini astronauts over Peru and adjacent areas in Bolivia and Chile have been compiled into semi-controlled photo image map at a scale of 1:1 000 000 (Figure 3) (MacKallor, 1968). This map was compiled by the Geological Survey in cooperation with the National Aeronautics and Space Administration, the Raytheon Corporation, and the Pan American Union. Individual photographs in the series were taken with the camera axis tilted as much as 50° from the vertical, but the significance of the resulting scale distortions was reduced in the laboratory by means of photogrammetric rectifier and ground control points obtained from reconnaissance scale base maps of the region. This photo image map, covering more than 250 000 square miles, is believed to be the first of its kind compiled from space photography and it demonstrates the feasibility of mapping large inaccessible areas from space in a short period of time. Relief displacement, a severe problem with the wide angle lenses used on conventional aerial cameras, is minimized on the rectified orbital photographs. This is because the orbital photographs are taken with a relatively narrow angle lens, and relief differences are small in comparison to the orbital altitude which is in excess of 100 nautical miles. Image position along join lines of adjacent photographs in the series is excellent.

The photo image map of Peru shows that photographs taken from orbital altitude have orthographic qualities not realized on conventional aerial photographs. In addition to geologic detail that may be interpretable, orbital photographs of somewhat larger scale than the Peru photos could be of direct use to the geologist as base maps, particularly in less developed areas where adequate base maps have not yet been compiled, or as a supplement to existing conventional base maps.

Geologic and physiographic features that have been identified on the photo image map include outcrop patterns of both intrusive and folded sedimentary rocks, volcanic features, landslides, intermittent lakes, and salt flats. The value of the synoptic view of a large area is exemplified on the mosaic by the prominent lineament which extends for several hundred miles across several photographs from the coastal area of southern Peru northeastward into the Andes. This lineament could be interpreted as the physiographic manifestation of a major fault, although it is not shown on the 1:1 000 000 scale geologic map of Peru and it is not known whether evidence of the lineament has been observed on the ground or on aerial photographs.

Also of interest are ring structures in the Andean Cordillera northwest of Lake Titicaca, previously not known, but recognized on the mosaic by several geologists from the Geological Survey, National Aeronautics and Space Administration, and private industry. Some of the structures are nearly 50 miles in diameter, but their geologic significance is not yet understood.

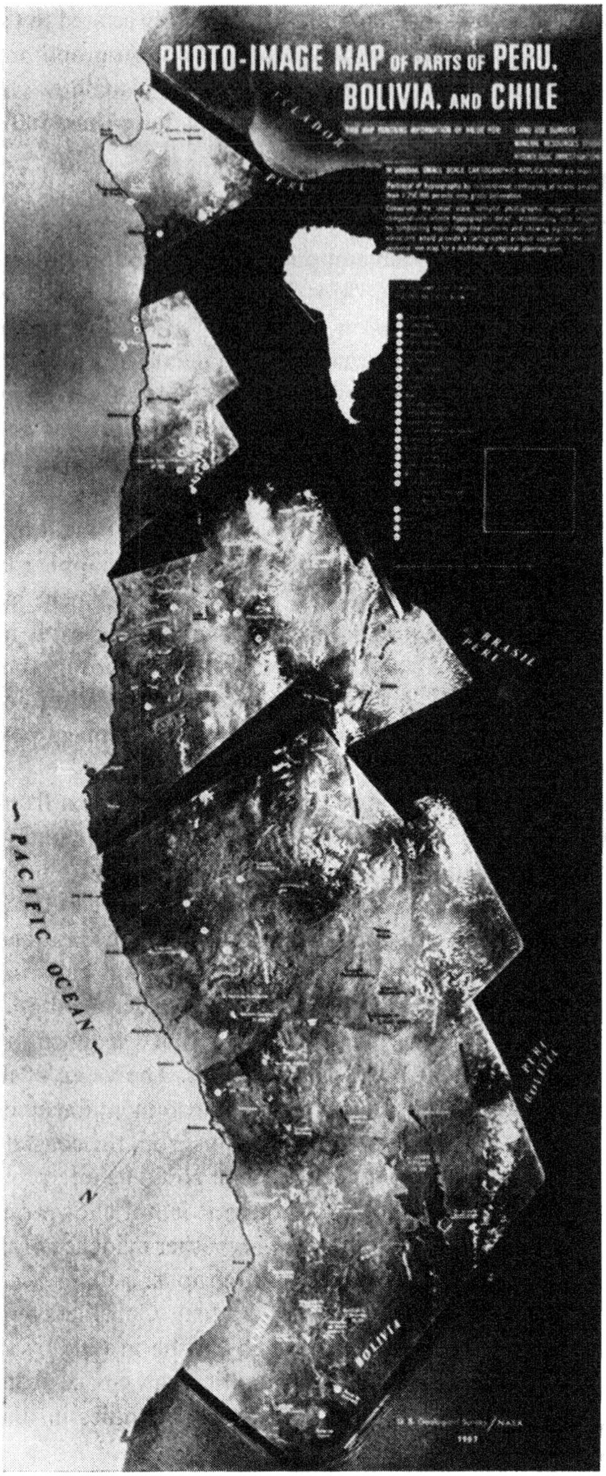
PHOTO-IMAGE MAP OF PARTS OF PERU, BOLIVIA, AND CHILE
ECUADOR
PERU
BRASIL
PERU
PACIFIC OCEAN
PERU
BOLIVIA
BOLIVIA
N

Other features readily discernible on the photo image map include mining areas, roads, irrigated areas, jungle agricultural sites, grasslands, snow fields, and even smoke from burning brush. Lima is not directly visible, although the metropolitan area is marked by a smog cloud on the color transparencies.

Comparison of the photo image map and maps compiled by conventional methods show substantial discrepancies between positions of points on the Pacific Coast and the shoreline of Lake Titicaca. A probable explanation of these discrepancies, however, is that although control points in a triangulation net are shown with great accuracy on conventionally compiled base maps, detail between control points is commonly less reliable. It is believed, therefore, that positions of some features may be more reliably shown on a large area photo image map than on the conventionally compiled base map.

3. Earth Resources Observational Satellite (EROS)

A. RATIONALE AND PROGRAM ORGANIZATION

The Earth Resources Observation Satellite (EROS) program of the Department of the Interior was conceived in 1966 largely as a direct result of the demonstrated utility of Mercury and Gemini orbital photographs to earth resource studies. The program is administered by the Geological Survey for the Department of the Interior in cooperation with other agencies, including the National Aeronautics and Space Administration (NASA) and the Department of Agriculture. Major objectives of the program are as follows:

(a) To design specific spacecraft data-collection systems on the basis of specific data user requirements.

(b) To disseminate data collected by satellites to appropriate scientists in order to facilitate assessment of land and water resources of the United States and other nations desirous of this information.

(c) To follow with modified satellites and improved instrumentation as required by operational needs of data users.

B. OPERATIONAL PARAMETERS FOR THE FIRST EXPERIMENTAL-PROTOTYPE SATELLITE

The first earth resources satellite is to be primarily an imaging satellite and will be designed for a useful life of one year. Orbital inclination is to be 93° to the equator at an orbital altitude of 500 nautical miles. The orbit will be sun synchronous at about two and a half hours before local noon, thus permitting nearly constant lighting and shadow conditions on successive passes over the same area every 17 days. Two vidicon cameras will permit simultaneous imaging in two spectral bands approximately centered at 5300 Å in the blue-green and 7500 Å in the red-infrared with bandwidths

---

Fig. 3. Photo image map compiled from Gemini photographs showing parts of Peru, Chile and Bolivia. (Prepared by the U.S. Geological Survey in cooperation with the National Aeronautics and Space Administration, Raytheon Corporation, and the Pan American Union.)

of 1000 Å and 1500 Å respectively. Because water achieves maximum transparency in the blue-green, but is relatively opaque in the red-infrared, this combination should be most useful both in imaging shallow sub-aqueous features, and in contrasting water areas against surrounding terrain. In addition, plants and many geologic materials exhibit marked contrasts in these bands. A third camera imaging in a spectral band centered at 6500 Å will aid in crop identification.

Cameras will cover a coincident ground area 100 miles on a side; overlap and sidelap of sequential exposures in the same and adjacent flight lines will permit mosaicking. Ground resolution desired is 100 to 200 feet. Area of each frame will be about 10 000 square miles.

C. BENEFITS TO THE GEOLOGIST AND HYDROLOGIST

Inasmuch as the first satellite is to be primarily an imaging satellite, probably the most significant direct benefit to the geologist and hydrologist will be a substitute for conventional base maps in areas where no coverage exists or where existing base maps are inadequate or obsolete.

Conventional methods of compiling reconnaissance scale base maps are time-consuming because they involve selection, interpretation, and compilation of pertinent data from a large number of aerial photographs and other sources. The photographs used for these studies may be months or years old; consequently the maps produced from them are of limited value in providing immediate assessment of damage and remedial measures in the wake of a flood or earthquake, and in study of these processes and manifestations of change with time.

However, terrain imaged from satellites will be nearly orthographic, and telemetered imagery reprints can be mosaicked with only the minimal rectification required to compensate for minor scale changes between frames. Because a single frame will include a ground area of about 100 miles on a side, the mosaic will require only a small number of horizontal control points from an existing base map or ground survey. More than 3000 conventional air photographs at a scale of 1:20 000 would be needed to cover one satellite image frame.

Earth resources satellite imagery will permit significant improvement in the efficiency of reconnaissance-scale geologic mapping. These improvements are estimated to range from 30 to 50% in areas where base maps are currently available, to 60 to 75% in areas where neither aerial photographs nor adequate base maps are available.

The broad overview of 'space' television imagery is expected to reveal major faults, folds, and physiographic features in a regional or continental context, and to permit studies of their interrelationships never before possible using conventional aerial photography. Repetitive coverage at nearly constant sun angle will also permit geologists to compare time variant phenomena such as delta formation, beach aggradation and erosion, sand dune development, floods, and terrain modifications due to volcanic eruptions and earthquakes.

In addition to facilitating geologic base map and interpretation problems, the satellite could serve as a communications relay for monitoring geophysical instruments installed in key areas. This use of the satellite is expected to yield data that will speed interpretation of the images and in time may provide substantial improvement over current methods for monitoring tilt meters, strain gauges, and seismometers located on the ground in key volcanic or earthquake areas. For the first time, it may become possible, for example, to develop an integrated world wide volcano monitoring network.

Many areas prone to instability are remote, thus complicating the use of conventional telephone landlines, which tend to be unreliable at the time of disaster. It is believed that a satellite communications system could alleviate some of these telemetry problems, although problems of data storage and effectiveness of telemetry for specific instruments must await further study by communications engineers. Jahns (1968) believes we are within a few years of having a warning system permitting advance notice of strong shocks along major faults in the United States.

D. GEOLOGIC AND HYDROLOGIC APPLICATIONS OF INFRARED AND RADAR IMAGERY

Infrared and radar imagery acquired from aircraft are being used to evaluate the applications of these sensors to earth resource studies to identify the possible utility of employing these sensors from orbit. Results to date are encouraging.

An airborne infrared imaging system uses a cryogenically cooled crystal to detect infrared energy emitted by terrain features in a swath beneath the aircraft. Mirrors focus the infrared energy on the detector and permit all elements in the swath-field to be scanned sequentially. Spatial positions of the terrain features are printed out in a two-dimensional plot on film in which the differences in intensity of the emitted energy appear in varying levels of gray; these differences can be related to the apparent temperature of the material.

Aircraft infrared imagery in the 8–13 micron band could show the distribution of thermal anomalies in an area. Such information could be useful in searching for new sources of thermal power or monitoring volcanoes.

Figure 4 is aircraft infrared imagery in the 8–13 micron band of a coastal area in the Philippines. The infrared imagery shows the dark toned or cool night-time surface of the land, and the generally lighter toned warmer water of Balayan Bay. Dark-toned lobate patterns along the shore are caused by discharge of colder ground water into the warmer waters of the bay, and it illustrates how infrared methods may be used to locate new sources of fresh water along coastal areas.

A future earth orbital IR imaging system could be used to document changes in temperature of large volcanic thermal targets, to map active lava flows at inaccessible eruption sites, and to supplement the International Volcanic Association's effort to document all volcanic eruptions. Potential hydrologic applications of infrared imagery could include the location of ground water discharges into large bodies of water. Another application would be to monitor the tidal and seasonal fluctuations of the so-called 'salt water wedge' in the estuaries of large rivers. Such information could be of use in evaluating the location of possible sources of fresh water in the estuary, and in assessing the susceptibility of the estuary to pollution.

An imaging radar consists of an airborne transmitter and receiver which operate in the millimeter-centimeter region of the electromagnetic spectrum. Gray-scale of the two-dimensional image is dependent upon the intensity of the reflected beam, which varies with the texture and dielectric properties of the reflecting surface.

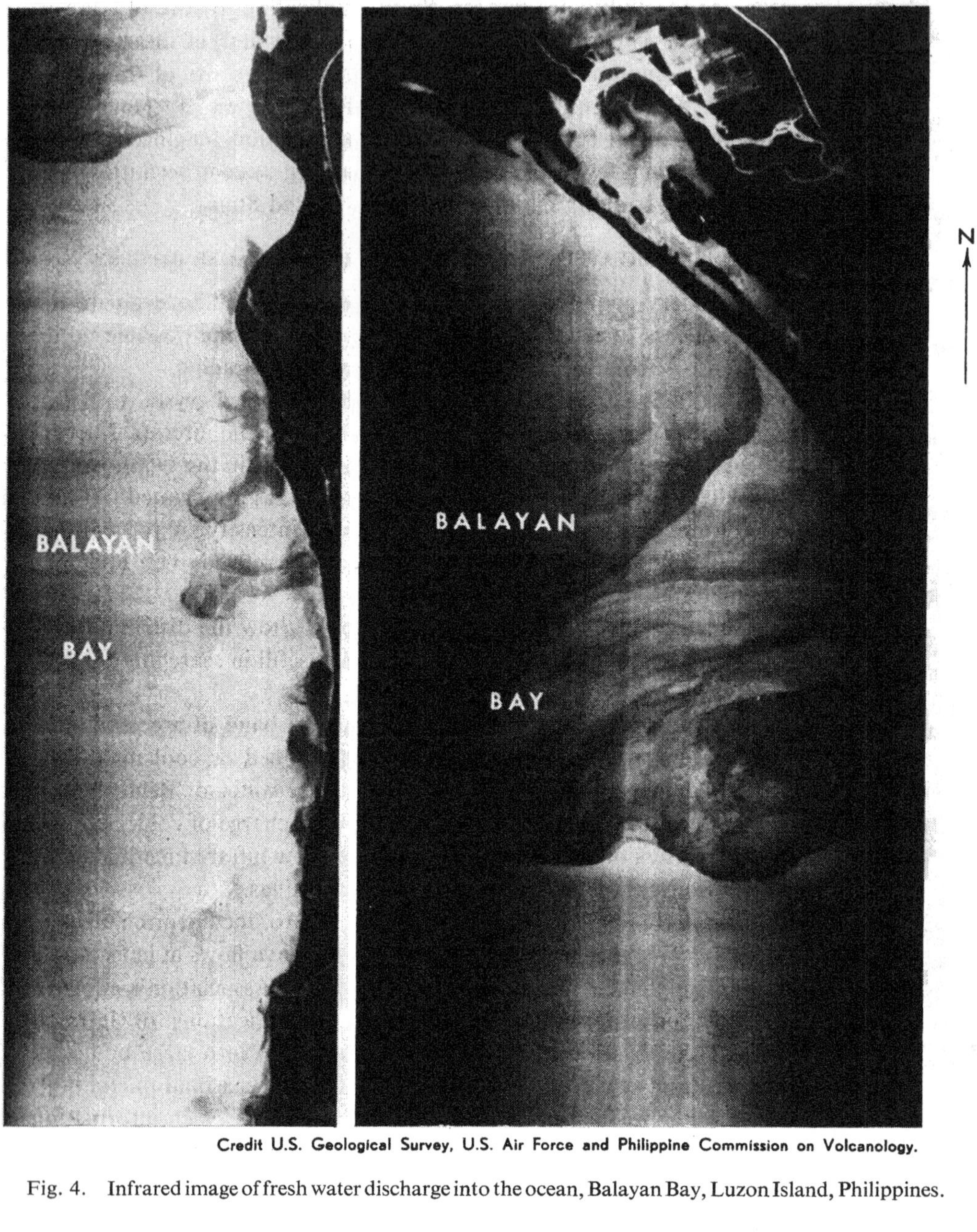

Credit U.S. Geological Survey, U.S. Air Force and Philippine Commission on Volcanology.

Fig. 4. Infrared image of fresh water discharge into the ocean, Balayan Bay, Luzon Island, Philippines.

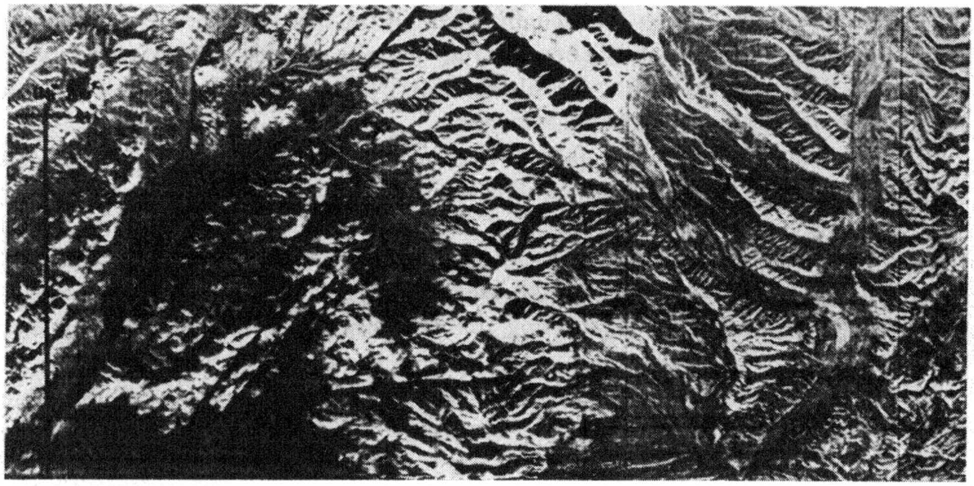

Fig. 5. Radar imagery of the Carlin gold mine area, Nevada. Major faults (dashed lines) show clearly on the imagery, but were overlooked in previous field studies and aerial photographs of large scale.

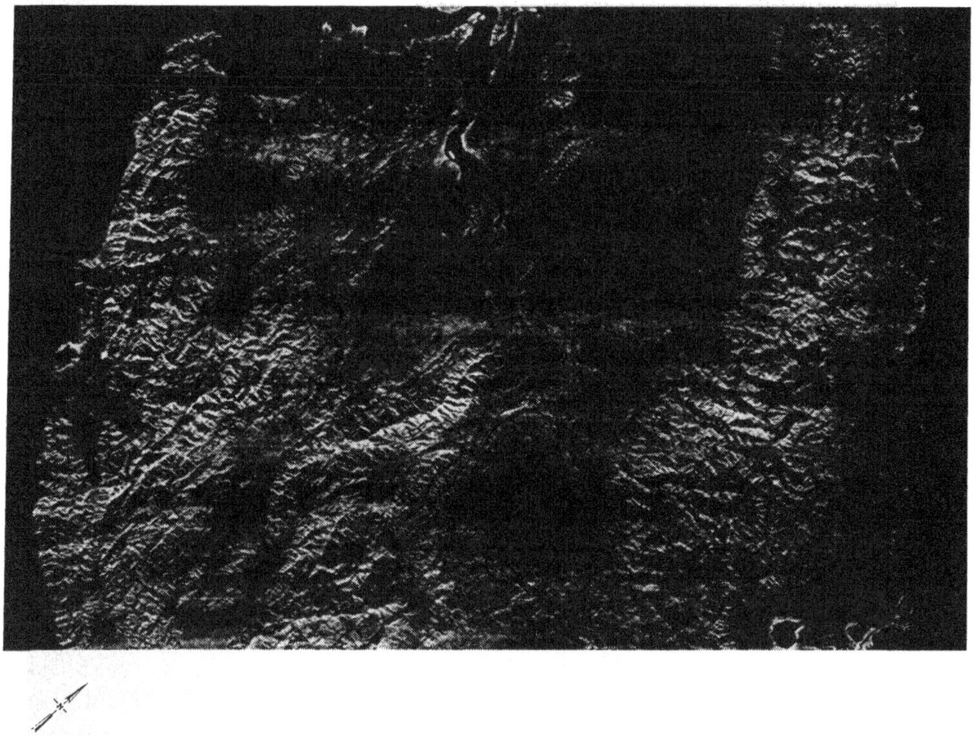

Fig. 6. Radar mosaic of about 8000 square miles of Darien Province, Republic of Panama and northwest Colombia. Despite persistent cloud cover and extensive jungle, the radar imagery reveals important information about topography and geologic structure of the area. (Courtesy of Westinghouse Aerospace, Raytheon Corporation, and the U.S. Army Corps of Engineers.)

Possible applications of long wavelength radars in the meter region are particularly exciting in earth resource studies because they may be capable of penetrating vegetation and thin layers of soil to reveal the underlying rock. These long-wavelength radars have not yet been evaluated, however.

Figure 5 is an aircraft radar image of the Carlin gold mine area in Nevada. This image reveals a series of northeast-trending faults that had not been detected by earlier field mapping or on aerial photographs of the area. Because these and similar structures are commonly important in the localization of ground-water and of mineral-bearing fluids that form ore deposits, knowledge of their location and orientation can be of direct importance in prospecting for new sources of minerals and ground water.

Figure 6 is an aircraft radar mosaic covering about 8000 square miles in jungle covered Darien Province, Panama and northwest Colombia. This mosaic was prepared by Westinghouse and Raytheon Corporations for the U.S. Army Corps of Engineers. Structural 'grain' of the region is clearly visible and attitude of folded rocks can be observed, particularly in the central part of the mosaic. The illustration is especially significant because this region is persistently covered by clouds and photography by conventional methods is rarely possible. Radar could be of use in other areas covered by persistent clouds such as parts of Guyana, Tierra del Fuego, the rain forests of central Africa, and the mountainous regions of southeast Asia.

4. Conclusion

Analyses of Gemini photographs illustrate several qualities, unique to orbital photography, that are of importance to geologists, hydrologists, and other scientists in assessing natural resources. These qualities of orbital photographs include the synoptic overview of a large area within a single exposure frame, water penetration superior to that achieved on conventional aerial photographs, and near orthographic positioning of topographic and cultural features.

The Earth Resources Observation Satellite (EROS) program was conceived in 1966 largely as a direct result of the demonstrated utility of Gemini orbital photography to earth resources studies. The first satellite is proposed to have a useful life of one year, and could be considered to be a prototype in a continuing operational program.

The first earth resources satellite would be primarily an imaging satellite, and it is expected that a major benefit to the geologist will be use of the high resolution television imagery as base maps in regions that are inadequately mapped or where dynamic changes have taken place. The broad overview of the imagery will also reveal physiographic features and evidence of geologic structure in a regional and continental context never before possible. Because the orbit will be sun synchronous, repetitive coverage at nearly constant sun angle will permit geologists to compare time-variant phenomena, such as erosion and aggradation of coastal areas, snow fields, and the terrain modifications due to flood, volcanic eruption, and earthquake. The satellite could serve also as a satellite communications relay for monitoring geophysical instruments located in key areas of potential geologic hazards.

Infrared and radar imagery acquired from aircraft are being used to evaluate potential applications of these sensors to earth resource studies. Results are encouraging and indicate that an orbital infrared imaging system would be useful for monitoring volcanic activity and surface temperatures of large bodies of water, and for other resource applications. Results of research also suggest that some imaging radar systems would also be useful for distinguishing between rock types, and for penetrating vegetation and thin layers of soil to reveal underlying bedrock.

References

Hemphill, W. R.: 1958, 'Small-Scale Photographs in Photogeologic Interpretation', *Photogramm. Eng.* **26**, 562–567.

Hemphill, W. R. and Danilchik, W.: 1968, 'Geologic Interpretation of a Gemini Photograph', *Photogramm. Eng.* **34**, 150–154.

Jahns, R.: 1968, Stanford University, in *Industrial Research*, p. 25 (Jan.).

MacKallor, J. A.: 1968, *A Photomosaic of Western Peru from Gemini Photography*, U.S. Geol. Survey Prof. Paper 600-C, p. C 169 – C 173.

ASTRONOMICAL RESEARCH WITH A LARGE ORBITING TELESCOPE*

LYMAN SPITZER, JR.**

Princeton University Observatory, Princeton, N.J., U.S.A.

Abstract. Research with an orbiting telescope of large diameter, with an aperture of 120 inches or more, would be of unique importance for solving some of the most important problems in modern astronomy. The performance characteristics of such a telescope will be briefly reviewed, together with some discussion of the relative advantages of manned and unmanned operation. Some of the research areas that would particularly benefit from such a powerful instrument will be discussed, including, in particular:

(1) The large-scale characteristics of the universe, including its geometry, homogeneity, size and curvature.

(2) The structure of galaxies and their evolution in time.

(3) The density distribution, chemical composition and physical state of the interstellar gas and the role of this gas in the formation of new stars and in the evolution of our galaxy.

(4) The density and temperature of the gas in the galactic halo or corona, which may originate from in-falling intergalactic material or from gas shot out of the nucleus of our own galaxy.

(5) Continuing and systematic study of planetary atmospheres, including changes of structure with position and with time.

(6) The surface detail of the planet Mercury and its implications for the early history of the solar system.

1. Introduction

The long-range research potentialities of large optical telescopes in space were considered by a group of astronomers in 1965 at the Woods Hole Summer Study, organized by the Space Science Board of the National Academy of Science.*** A central conclusion of the Working Group on Optical Astronomy was as follows:

> We conclude that a space telescope of very large diameter, with a resolution corresponding to an aperture of at least 120 inches, detecting radiation between 800 Å and 1 mm, and requiring the capability of man in space, is becoming technically feasible, and will be uniquely important to the solution of the central astronomical problems of our era.

Essentially this same group† was asked by the Space Science Board in 1966 to constitute a committee of the Board for continuing studies of this powerful instrument, which has come to be referred to as the Large Space Telescope (LST). This LST Committee has devoted most of its efforts to a detailed consideration of the astro-

* Adapted from a paper in *Science* **161** (1968), 225–229.

** The author is Charles A. Young Professor of Astronomy at Princeton University, Princeton, N.J., and director of the Princeton University Observatory.

*** *Space Research, Directions for the Future* (Space Science Board, National Academy of Sciences National Research Council, Washington, D.C., 1966), pt. 2.

† Members of the LST Committee are as follows: A. D. Code (University of Wisconsin); L. W. Fredrick (University of Virginia); F. J. Low (University of Arizona); G. Münch (Mt. Wilson and Palomar Observatories); H. J. Smith (University of Texas); L. Spitzer (Princeton University), chairman; W. G. Tifft (University of Arizona); and F. L. Whipple (Smithsonian Astrophysical Observatory).

nomical research that could be carried out with such an instrument. A series of conferences were held with other interested astronomers, and a report, 'The Scientific Uses of a Large Space Telescope', has been prepared. Here I summarize the salient features of this report.*

2. The Role of Man

Observations with large optical telescopes in space, designed to detect and analyze the faint light from remote stars, nebulas, and galaxies, promise to have a revolutionary impact on astronomical research. Because of the size and complexity of the necessary equipment, this field has developed slowly. Even with small telescopes on sounding rockets, precise pointing toward stars other than the sun is technically difficult, and not until 1965 was this stabilization sufficiently precise to yield ultraviolet stellar spectra with enough resolution to show the strong absorption lines. In principle, however, the ability to make observations from above the atmosphere should greatly increase the effectiveness of an optical telescope; the absence of an atmosphere, with its strong absorption at wavelengths shorter than 3000 Å and longer than 10 000 Å, permits extension of the wavelength range into the ultraviolet and the infrared, and the absence of scintillation and image motion caused by atmospheric irregularities makes it possible, in principle, to obtain diffraction-limited imagery, with a resolution limited only by the aperture of the telescope and the wavelength of light. While in ground observation, high resolution may be achieved for bright stars and planets through the use of various sophisticated techniques, there appears little prospect that this objective can be achieved for faint objects, which must be observed for long periods of time with telescopes of large aperture.

The role of man in connection with a large astronomical observatory can be a very important one. A large facility of this type may be visualized as having a life of many years, or even decades. It is very doubtful whether such a complex instrument could function effectively for so long a period without occasional maintenance and repair. Moreover, as new types of instrumentation become developed, it would be desirable to replace and update the detectors, image tubes, spectrometers and other types of equipment that are used to analyze the light reaching the focal plane of the telescope. For both these functions – maintenance and updating – man in space would seem required. Initial adjustment and testing of delicate optical equipment is probably also best carried out by man. One may visualize a small team of men visiting such a telescope at intervals, replacing faulty modules, inserting improved instrumentation, and checking the operation before returning to Earth.

The advantages of continuous manning for a space telescope are much more controversial. Even with telescopes on the ground, the present tendency is more and more towards automatic operation, with exposure meters, automatic guidance, etc. Out in space, where manpower will be scarce and expensive, this tendency is likely to be enhanced. Since the telescope and its instrumentation must be *in vacuo*, any astronaut

* Much of the concluding section of this article is taken directly from the committee report.

present during operation of the telescope would have to be in a space suit, severely limiting his effectiveness. Furthermore, for some measures the perturbations produced by the presence of man might be a disadvantage, making precise guidance very difficult. We may conclude that any space telescope should be operable in a fully automatic mode, with ground control of the detailed observational program.

One function that might require relatively frequent visits by man is the use of photographic film. Changing film cassettes and developing the exposed film might be difficult to automate. However, it seems unlikely that photography will be used very much in space observations, since television-type tubes offer the advantages of higher sensitivity, easier data handling and transmission, and freedom from problems of storage and of fogging by energetic protons and electrons.

The extent to which astronauts should be involved in a permanent space observatory will depend, of course, on economic factors. At present cost levels it would probably be less expensive to launch a new telescope of appreciable size every year or two than to send up teams of astronauts at intervals for repair, maintenance, and updating. However, if the cost of resupply of a space station is significantly reduced, maintaining a permanent facility with occasional visits becomes economically attractive, and could make a relatively larger space telescope feasible.

3. Performance of the Large Space Telescope

Before considering how astronomy could profit from use of the LST, let us review the technical performance characteristics that might be expected for such an instrument. Basically, a space telescope, like a telescope on the ground, is simply a large concave mirror that reflects light to a sharp focus, where it is analyzed by any of a variety of instruments, including precise photometers, spectrophotometers, polarimeters, and image recorders such as photographic film or a television camera. To achieve a long focal length, a conventional two-mirror system, of Cassegrain type, is likely to be used, with a secondary convex mirror reflecting light back through a hole in the primary mirror to a focal plane a short distance behind the primary. With an optimum choice of aspheric surfaces on the two mirrors, such a system can give diffraction-limited images over an angular field several minutes of arc in diameter. With aluminum coatings on the two mirrors, and with thin overcoats of lithium fluoride, reflection coefficients as great as 60% can be obtained at wavelengths as short as 1000 Å, and coefficients greater than 90%, at wavelengths in the visible and far-infrared regions of the spectrum. Thus a space telescope of this relatively conventional optical design could be used to concentrate electromagnetic radiation at wavelengths between 1000 Å and 1mm, the shortest radio wavelength readily accessible from the ground. Photons of wavelengths between 912 Å and at about 50 Å in the soft X-ray region, are strongly absorbed by interstellar hydrogen atoms.

For an ideal optical system, with a mirror of radius a, the first dark ring around the image of a point source, observed with light of wavelength λ, has an angular radius of 0.61 λ/a radians. For a 120-inch (3-meter) telescope and light of 5000-Å wavelength, this radius is 0.04 sec of arc; ideally, 84% of the light at this wavelength falls inside this circle. Actually, photographic images obtained with ground-based instruments normally have a radius about ten times this, or a diameter of about 1 sec of arc. Sharper images have been obtained with ground-based instruments either visually or with very short photographic exposures, but one can generally not count on these for systematic research programs on faint objects. The combination of high spatial resolution and wide wavelength coverage constitutes the essential advance in capability offered by the LST.

To realize and make effective use of this capability would require a large engineering effort. To achieve nearly the ideal resolution possible, the surfaces of the two mirrors should not deviate from their ideal shape by more than $\lambda/50$ root mean square; careful control of the thermal design would be required to achieve this objective. The pointing of the telescope should be held steady during an exposure of several hours, with angular deviations from the mean direction preferably not exceeding a tenth of the resolution, or 0.004 sec of arc for imagery at 5000 Å with a 120-inch telescope.

Studies for the National Aeronautics and Space Administration, together with test data and flight results obtained from the Stratoscope and Orbiting Astronomical Observatory programs, indicate that these objectives can be met with present techniques for an orbital telescope in the 40-inch range, and probably also for an instrument of roughly 120-inch aperture. The extensive equipment required can be visualized as a permanent space observatory. To ensure reliable operation, the installation would have to be manned at relatively frequent intervals. Discussion of the many engineering problems involved in designing and orbiting such an installation is beyond the scope of this article; it is assumed here that feasibility has been sufficiently indicated to warrant detailed discussion of ways in which the LST might be used scientifically.

The LST could be placed either in an orbit around the earth or on the surface of the moon. The cost of launching, maintaining, and resupplying such an installation would be least in an orbit relatively near the earth's surface at an altitude of some 500 km. In such an orbit most objects are occulted by the earth for an appreciable fraction of each 100-min orbital period. A higher orbit or a location on the moon would offer some advantages for more continuous observation, especially for longer exposures, but it is not clear whether these advantages would be worth the greatly increased cost.

One basic operating characteristic of any telescope is the faintness of a star that can be detected with a given exposure time. A fundamental limit is set by the number of photons available. From a star of stellar magnitude 24 (about the faintest so far detected from the ground), about 20 photons per sec, with wavelengths between 4500 and 5500 Å, strike a telescope mirror of 120-inch diameter. Some 10% of these can be reliably detected photoelectrically. Hence, with an exposure time of 2 hours, from a star 1/100 as bright as this (that is, of magnitude 29), 140 photon pulses would be counted, providing a detectable signal with a measurement accuracy of 10% (0.1 magnitude), if the background noise were sufficiently low. On the earth, the permanent airglow from the night sky, which is equivalent, under good conditions, to the brightness of one star of magnitude 22 per square sec of arc, masks such very faint stars, since the stellar images are about 1 square sec of arc in diameter. With a 120-inch space telescope the image area would be about 0.005 square sec of arc; since in space the sky is less bright by about one stellar magnitude (or a factor 0.4) than the sky as seen from the earth's surface, the diffuse sky light within an area equal to the stellar image would be about equal to the light from a 29th-magnitude star. Thus the limit imposed by the sky brightness is much less important in space than it is on the earth, thanks largely to the much sharper angular concentration of a stellar image obtainable in space. With a photoelectric image tube and either photographic recording or electrical recording as in the case of television cameras, stars as faint as magnitude 29 should be detectable by the LST, with exposure times of some 5 to 10 hours or less.

4. Galaxies and the Universe

While the LST would have important uses in many branches of astronomy, in the study of objects outside our own galaxy the capabilities of this instrument seem particularly well matched to the scientific requirements. The nature, extent, and destiny of the vast universe in which we live have engrossed man for thousands of years, but only now is there reason to hope for meaningful answers. In particular, bringing all galaxies in the universe effectively ten times closer than they can be brought by ground-based instruments promises to provide the information needed to answer some of these most basis questions in astronomy.

More specifically, research on other galaxies is designed to resolve the following two central problems:

(1) What are the large-scale characteristics of the universe – its geometry, homogeneity, size, and curvature? What have these characteristics been in the past? What will they be in the future?

(2) What is the structure of galaxies, and how do these systems evolve with time?

Investigations of these two central questions are somewhat intertwined, but the two areas of research are distinct in principle. In the first, the spatial distribution and velocities of the galaxies are investigated, without regard to their structure. In the second, the individual galaxies are analyzed. In both these investigations the high resolution and broad wavelength range of the LST could be of crucial importance.

The velocities of galaxies can be determined from the Doppler shifts of the lines in their spectra, which can be obtained with large instruments on the ground, such as the 200-inch Hale telescope on Mt. Palomar. To establish the distribution of galaxies in space, however, requires a knowledge of their distances, which are much more difficult to determine. In principle, the distances can be measured in two ways – either photometrically, from the apparent brightness of an object whose absolute luminosity is known, or geometrically, from the angle subtended by an object of known size. Most distance determinations of galaxies have been based on the photometric method, the brightness of individual stars being used for very close galaxies and the brightness of an entire galaxy, for more remote systems. The absolute luminosities of some individual stars can be determined rather accurately, especially in the case of the brilliant Cepheid variable stars for which the luminosity is strongly correlated with the period of variation. Hence distances determined from stellar photometry are relatively trustworthy. Galactic luminosities, on the other hand, are difficult to determine, and distances determined in this way may be subject to substantial errors.

One powerful capability of the LST would be that it could detect Cepheid variables, and measure their luminosity, at a distance of 10^8 light-years, 10 times the distance at which such variable stars can be measured from the ground. At this distance of 10^8 light-years the velocity due to the expansion of the universe is believed to be roughly 3000 km per sec, sufficiently large to be relatively unaffected by random galactic motions. Thus the Hubble constant, H (the ratio between the expansion velocity and the distance) could be determined very precisely with the LST. The greatest distance at which Cepheid variables can be measured from the ground is not much more than 10^7 light years, and at this distance the expansion of the universe is largely masked by random local velocities. According to the 'big-bang' theory, the time $1/H$ is proportional to the time since the 'creation' or starting point of the expansion. Since the constant of proportionality in this relation depends on the structure of the universe, comparison of a precise value of $1/H$ and the ages of the stars should give basic information about the nature of the universe.

Distance measures for galaxies farther away than 5×10^9 light-years would give the large-scale curvature of the universe. In particular, non-linearity in the velocity-distance relationship at great distances depends on the radius of curvature. The LST could be of vital importance in determining the distances of the most remote galaxies. The accuracy of values obtained photometrically would be improved by measurement at infrared wavelengths, since measurement in the visible region of the spectrum involves uncertain corrections for the red shift that results from the large velocity due to expansion at great distances. Also important would be the ability to obtain measures of the angular diameters of galaxies from detailed luminosity profiles. The turbulence of the atmosphere, with resulting fluctuations in the seeing, and fluctuations of the airglow make it quite impossible to obtain such measures from the ground for galaxies some 5×10^9 light-years away (angular size, at most 1 sec of arc), and complicate such measurements for closer systems. The higher resolution and fainter, more stable brightness of the sky background in a space telescope should make measurement of angular diameters from the LST quite feasible. Similarly, for systems within about 10^9 light-years, the angular diameters of hydrogen-emitting regions, which would be measurable with the LST, could provide a good determination of distance, with probably much less scatter than is associated with photometric measurement of distances of individual galaxies.

Such considerations suggest a fascinating possibility. If the universe has an observably small radius of curvature, say about 5×10^9 light-years, the apparent diameters of objects at much greater distances should be observed to increase with increasing distance. Thus the plot of apparent diameter versus red shift should have a minimum, reached at a red shift, $\Delta\lambda/\lambda$, which depends on the average density of matter in the universe, and which is about unity if the gravitational energy is equal, and opposite, to the kinetic energy of expansion. A measurement of this effect would provide convincing proof of the overall curvature of the universe and would give directly values for both the radius of curvature and the corresponding density of matter.

Research on the second problem, the structure and evolution of galactic systems, would evidently be much advanced by the resolution capability of the LST. For example, compact nuclei are believed to be responsible for much of the explosive activity observed in galaxies, including the high luminosities observed from quasars, the high excitation observed in Seyfert galaxies, and the outward-streaming filaments of gas observed in a variety of systems. At least in the closer systems, the structure of galactic nuclei should be observable with the LST. A system 1 light-year in radius could be at least partially resolved even at a distance of 3×10^6 light-years; several dozen galaxies lie within about this distance. Some of the most active galactic nuclei are apparently much more luminous in the far-infrared, at wavelengths longer than 10 microns, than in the visible spectrum, and for detection and spectroscopic study of such systems even at very great distances the LST could be particularly useful.

Evolution of galaxies through time should produce measurable differences between systems some 5×10^9 light-years away and those close at hand, since the light from these remote systems started out toward the earth some 5×10^9 years ago, when the universe was perhaps half as old as it is now. The LST would permit examination of the structure of these remote systems, which, on pictures taken from the ground, appear only as tiny formless blobs.

Research on galaxies with the LST would not be a quick program, especially since exposures of several hours on many different objects would be required; it would be several years at least before any conclusive results could be obtained. However, it seems clear that the amount of detail visible in these stellar systems could be vastly increased with such an instrument. Surprises would doubtless turn up, perhaps fundamentally altering our viewpoint, but even the foreseeable results promise to add very greatly to our general knowledge of the universe.

5. Gas between the Stars

Interstellar matter is believed to play a major role in the evolution of a galactic system. Groups of new stars form from interstellar clouds, and massive old stars, in which heavy elements have been produced, explode as supernovae, ejecting their material back into interstellar space.

To study these phenomena the density, velocity, composition, and physical state of the interstellar gas must be determined. Important results have been obtained from measures of the radiofrequency line at 21-centimeter wavelength produced by neutral hydrogen atoms, and, in the visible spectrum, of absorption lines produced by neutral sodium (the D lines at 5890 and 5896 Å) and ionized calcium (the K and H lines at

3934 and 3969 Å). However, these lines are relatively weak; the oscillator strength for the 21-centimeter line is less by a factor of 10^{-12} than that for normal, 'permitted' lines in the visible spectrum, and neutral Na and singly ionized Ca^+ are both relatively scarce, since the more highly ionized Na^+ and Ca^{++} predominate over neutral Na and singly ionized Ca^+, because of the very low electron density in interstellar space. The ultraviolet absorption lines of the abundant elements are very much stronger. For example, the O I line at 1302 Å will have an easily measurable equivalent width if the chemical composition of the interstellar gas is similar to that of the sun, and if the amount of interstellar gas in the line of sight is some 10^{-8} g per cm^2, which is about 10^{-3} as much as the amount needed to produce either observable absorption in the Na and Ca^+ lines or detectable emission in the 21-centimeter line of hydrogen. Most elements produce no interstellar absorption lines in the visible spectrum, since almost all atoms are in the ground level in interstellar space and a photon energy of more than 4 eV (corresponding to a wavelength of less than 3000 Å) is needed to excite atoms in permitted transition from the ground level. Hence the interstellar gas can be detected and analyzed from above the atmosphere with sensitivity greater by three orders of magnitude than that for ground-based observations.

To some extent this ability to detect the presence of very small amounts of interstellar gas can be exploited by smaller space telescopes. However, the LST, with its large photon-collecting area, could undertake research programs which are not feasible with smaller instruments. With a photoelectric image tube and exposure time of about an hour, a 120-inch telescope could obtain an ultraviolet spectrum of a 10th-magnitude blue star with the resolution of 0.1 Å needed for weak interstellar lines. Such a star would be at a distance of several thousand light years from the earth, the exact distance depending both on the absolute stellar luminosity and on the amount of obscuration produced by small solid particles or dust grains in the line of sight. Measures on such distant stars could yield far more detail on the galactic distribution and dynamics of the interstellar gas than can be obtained in any other way.

One of the significant problems that could be investigated with such equipment is the galactic corona or halo, the envelope of gas that extends thousands of light-years above and below the galactic plane. Present information on this subject is fragmentary; a few high-velocity clouds have been observed, from weak 21-centimeter emission and from Ca^+ line absorption. This gas may originate from intergalactic material falling into our own system or from gas shot out of the nucleus of our galaxy. Confinement of cosmic rays in this galactic halo has been suggested, but this possibility remains speculative for lack of data. Measurement on such physical quantities as the ionization and temperature of the gas, which are believed to be profoundly affected by the energetic particles in the cosmic radiation, and the detailed velocity distribution of different clouds of gas would open up essentially a new field of research.

The motions of gas in spiral arms, which may have an important effect on the structure of these features, could also be investigated with the LST. An intrinsically bright star with an apparent magnitude of 10, the limiting magnitude, as noted above, for spectroscopic observation with the LST, might be some 10 000 light-years away. This distance is great enough so that stars in other spiral arms, in addition to the local arm in which the sun is situated, could be observed with the LST, and the gas in these arms could be analyzed in detail. One outstanding problem, which is evidently vital to our understanding of the way in which spiral arms form and are maintained, is the ratio of gas densities between and inside the arms; whether this ratio is as high as a few percent or possibly very much less would perhaps be revealed by LST observations.

The infrared capabilities of the LST could also contribute significantly to research on interstellar matter. Extended infrared sources or infrared nebulas have been observed through available infrared windows in our atmosphere. Some of these sources are believed to be contracting interstellar clouds about to condense into a multitude of new stars. Both the resolution capability of the LST, which would give a beam of 1 sec of arc at a wavelength of 10 microns, and the large collecting area for photons, which would facilitate detailed spectrophotometry, should be useful in analyzing the way in which stars form from interstellar clouds.

6. Solar System

For systematic study of the planets and their satellites the LST would be a most useful instrument. Even for Mars and Venus, the two planets that have been reached by planetary probes, a 120-inch telescope in orbit around the earth could yield systematic and continuing data not obtainable in any other way, except with special observatories orbiting around the planet of interest. The telescope's resolving power of 0.04 sec of arc would give resolution of surface detail ranging in size from 8 km on Venus to about 700 km for Uranus and Neptune.

Continuing observations of the cloud systems on these planets could open up a new field of research in planetary weather. This possibility would be enhanced by spectroscopic observations with high frequency resolution and high spatial resolution as well. For example, variation of the planetary radiation, including various absorption lines, very near the edge of the planetary disk would give detailed information on the distribution, with height, of various atmospheric constituents. In addition, the temperature of the Martian atmosphere, as a function of position and height, could be determined from emission profiles of molecular lines in the infrared, such as the 15-micron band of CO_2, with high spatial and spectral resolution. Changes of these distributions with time, together with changes in apparent cloud structure shown in high-resolution images, would provide observational data needed for a detailed analysis of planetary meteorology.

The surface detail on Mercury would be of particular interest. If the history of that planet has been similar to the history of the moon, several hundred craters should be detectable on Mercury with the LST. The number of craters observed might give important information on the early history of the solar system, especially on the number of meteors and asteroids in the solar system relatively close to the sun. In order to photograph a planet so close to the sun without excessive heating of the instrument, special arrangements would probably be required to keep the LST in shadow during the observations. Surface details on the larger asteroids and on the satellites of Jupiter would also be interesting.

7. Conclusions

The programs outlined above represent some of the exciting astronomical problems suggested for the LST. This great instrument could be of central importance not only in the study of cosmology and the other areas described above but in many additional areas of astronomy, including cometary research, studies of stellar evolution, and the determination of stellar masses and distances.

Experience indicates that some of the most exciting results obtained in a space program are likely to be those which had not been foreseen. Improvement of observational capability by an order of magnitude, both in spatial resolution and in photon-gathering capacity at ultraviolet and infrared wavelengths, is bound to lead to unexpected results in the study of the heavens.

While the LST would be an extremely powerful tool, its efficient utilization would require active astronomical programs both with other, smaller space instruments and with ground-based facilities. Orbiting telescopes with apertures of 30 to 60 inches and smaller instruments launched in sounding rockets can be used for a variety of important astronomical problems, such as determination of the ultraviolet luminosity of hot stars and the chemical composition of the interstellar gas within several hundred parsecs. Quite apart from the engineering desirability of launching smaller space telescopes before building the large instrument, the astronomical requirement for a continuing series of smaller space telescopes should be an overriding consideration in setting the pace of the LST effort.

Similarly, the LST must rely heavily on ground-based research both in selecting objects to observe and in interpreting the results. For example, the red shifts and other characteristics of different galaxies and the spectral types and colors of wide classes of stars would have to be observed from the ground before LST time could be allocated intelligently. Particularly for the southern hemisphere of the sky, such information is now fragmentary; if the LST were available today, an appreciable fraction of its time would probably be used in making observations of southern stars and galaxies, some of which could be made about as well (and much more economically) from the ground as from space. Moreover, for spectroscopic observations in visible light, a large instrument on the earth's surface is generally preferable to a space telescope. Ground-based astronomy, including research with X-rays, γ-rays, and radio waves as well as with solar, stellar, and nebular light, forms a single discipline, and will best advance together. A proper balance between the various parts of this unified discipline should be provided, both in scientific effort and in funding. For a small percentage of the cost of the LST program, the facilities for ground-based astronomy could be doubled. Such a step would increase both the value and the efficiency of the space astronomy program.

In summary, the following four conclusions may be drawn:

(1) The LST would make a dominant contribution to our knowledge of cosmology – to our understanding of the content, structure, scale, and evolution of the universe.

(2) In many other fields of astronomy, also, the LST would give important and decisive information.

(3) An efficient space astronomy program cannot be carried out by the LST alone. A continuing series of smaller space telescopes is also required.

(4) The most effective utilization of powerful space telescopes requires a substantial increase in the number of ground-based instruments.

Bibliography

Other publications by Lyman Spitzer, Jr. on Space Telescopes:

1956, 'Astrophysical Research with an Artificial Satellite', Franklin Institute of Pennsylvania Monograph No. 2, pp. 69–78.

1959, 'Interstellar Research with a Spectroscopic Satellite' (with F. Zabriskie), *Publ. Astron. Soc. Pacific* **71**, 412–420.

1960, 'Space Telescopes and Components', *Astron. J.* **65**, 242–263.

1962, 'The Beginnings and Future of Space Astronomy', *American Scientist* **50**, 473–484.

ORBITAL LABORATORY – STEPPING STONE TO INTERPLANETARY FLIGHT

WALTER M. HOLLISTER

Dept. of Aeronautics and Astronautics, Massachusetts Institute of Technology, Cambridge, Mass., U.S.A.

Abstract. An orbital laboratory is a necessary link in a manned spaceflight program leading to planetary exploration. Missions to the planets call for total flight times in excess of one year, while manned flights to date have been limited to total flight times of two weeks or less. Almost all of the technical and operational problems imposed by manned planetary flight, over and above those for lunar flight, are directly related to the major increase in mission duration. Experimental data on the exposure of humans to weightlessness is needed for these longer time periods to establish the medically and operationally acceptable levels of artificial gravity for interplanetary flight. Operating lifetimes for spacecraft subsystems need to be demonstrated. High reliability demands an onboard maintenance capability which will have to be proven feasible during earth orbit before being committed to an interplanetary trip. The main spacecraft of an interplanetary expedition performs essentially the same functions as an orbital laboratory. Each acts as a long-lifetime shelter, a laboratory, a supply depot, a docking station, and a repair facility. By proper choice of the interplanetary mission, the main spacecraft could actually be an orbital laboratory placed onto an interplanetary trajectory which makes a free-fall return to Earth after a flyby of one or more planets. It is possible to continue the interplanetary free-fall indefinitely with no propulsion required other than that needed to correct for guidance errors. A total, long-range program for manned spaceflight is outlined which proceeds from present accomplishments to a laboratory orbiting earth, thence to the simplest interplanetary flyby missions, on to the flyby/lander mode of interplanetary exploration, and ultimately to orbital laboratories on permanent free-fall trajectories between the planets. Each step of such a program would build logically on the accomplishments of the earlier missions. The development of an orbital laboratory as a durable, long-life facility is the key link in the overall program leading to manned planetary exploration.

1. Introduction

Ambitious planning took place in the early 1960's for manned missions to the planets which would start as early as 1970, using the technology developed for lunar flight. Those plans were optimistic. Lunar missions call for a total duration on the order of two weeks while missions to the planets call for flight times in excess of one year. The major technical and operational problems imposed by interplanetary flight over and above those for lunar flight are directly related to the increase in total mission time. Operational experience must be gained with long-duration missions near earth before setting off for the planets. The obvious way to gain that operational experience is with an orbital laboratory. The main spacecraft of an interplanetary expedition is essentially an orbital laboratory that has been placed in an orbit around the Sun instead of an orbit around the earth. The difference between the two is the significant time interval between subsequent opportunities for abort, rescue, or resupply from earth. A rescue vehicle can rendezvous with an orbital laboratory in a few hours [1]. An abort from an interplanetary mission can take from several months to the full length of the mission [2]. The functions of the orbital laboratory and the interplanetary

S. F. Singer (ed.), Manned Laboratories in Space.

spacecraft are the same. Each acts as a long-lifetime shelter, a laboratory, a supply depot, a docking station, and a repair facility. Operational experience in performing these functions should be gained with the orbital laboratory while demonstrating the necessary reliability. The training and practical engineering for interplanetary flight should be carried out simultaneously with the operation of the orbital laboratory.

A total, long-range program for interplanetary exploration should build logically on the accomplishments of earlier missions. By proper choice of the interplanetary mission, the main spacecraft could actually be an orbital laboratory placed onto an interplanetary trajectory which makes a free-fall return to earth after a flyby of one or more planets. It is possible to continue the interplanetary free-fall indefinitely with no propulsion required other than that needed to correct for guidance errors. This paper will outline a program which proceeds in logical steps from the development of an earth-orbital laboratory to a system of solar-orbital laboratories on permanent free-fall trajectories between the planets. The laboratories are durable, long-life facilities with the operational requirements similar throughout to those for the orbital laboratory. The suggested program therefore reinforces the necessity for development of the orbital laboratory as a step on the path leading to interplanetary flight.

2. Objectives of Interplanetary Flight

The primary purposes for manned, interplanetary flight are scientific research and exploration. Major objectives are to search for extra-terrestrial life, to determine the origin and evolution of life, to investigate the origin of the earth and the solar system, and to understand the physical processes that take place within the solar system. The first destinations will be the closest planets, Mars and Venus. There is a strong probability that simple life forms have developed in some variety on Mars and perhaps in the upper atmosphere of Venus. The simplest missions to these planets consist of a flyby where the objective is to pass close enough to the planet to make observations and gather scientific data. A more difficult mission is to go into orbit about the planet for a period of time prior to the return trip to earth. The most difficult mission is to perform a landing on the surface. The three classes of missions are called respectively flyby, orbiter, and lander. Studies of each class have been performed.

The individuals put into space under the manned space programs to date have been primarily test pilots. Eventually it will be primarily scientists who will go. Astronomers have through history often travelled long and difficult journeys in order to view celestial events. Eventually they will have the opportunity to go into earth orbit for improved viewing or on to the planets themselves. Missions to the more distant planets require longer flight times. Consideration will be given to *sustained* operations in space. There is the eventual possibility of man-and-wife teams in the crew, birth in orbit, and families who spend a major part of their lives in orbit. Scientists exploring the oceans have already set similar goals for sustained operations under water. There is speculation that persons might spend their entire lives working in the sea while living in shelters established on the ocean floor [3]. Similar long-range goals are reasonable for space. An orbital laboratory is the place to develop the capability for sustained operations in space.

3. Flyby Missions

The advantage of the flyby mission is that it is relatively simple. No propulsive thrust has to be applied to go into or out of a parking orbit at the destination. The spacecraft does not undergo large stress so that the overall design is simple. The spacecraft could be an orbiting laboratory put on an interplanetary trajectory. Additional advantages can be made of the flyby maneuver. During the close encounter the spacecraft is accelerated due to the presence of the gravitational field of the planet. Significant changes in the characteristics of the spacecraft orbit result from this acceleration, whereas the change in the planetary orbit is negligible due to the large planet-spacecraft mass ratio. As an example, it takes less propulsive energy and less time to make a round-trip flyby that encounters both Venus and Mars than it does to encounter only Mars. The idea of a double flyby was considered as early as 1925 by Hohmann [4]. An actual trajectory was suggested by Crocco [5] in 1956 and has been studied subsequently by others [6], [7], [8]. The largest velocity change possible as a result of a single flyby is equal to the value of circular satellite velocity at the planet's surface. This is 3.6 km/sec for Mars, 7.2 km/sec for Venus, and 7.9 km/sec for earth. In general, more advantage can be taken of a flyby of earth or Venus than of Mars. This is a partial explanation of why a flyby of Venus can be an advantage when en route to Mars.

The flyby mission enjoys a major advantage in reliability over the orbiter or lander missions. The spacecraft is always on a trajectory which will bring it back to the vicinity of earth. The orbiter and lander missions always face the possibility of a failure of the departure propulsion system after arrival at the planet. For these reasons it is often suggested that a flyby mission be planned as the first attempt to send men to the planets. It is a logical next step after the development of an orbital laboratory. The major new technical problems to be resolved are the addition of a propulsion system to launch the spaceship from earth orbit initially and a re-entry capsule to recover the crew at the end of the mission. The module containing the radiation shielding, the shelter, and life support for the crew would be abandoned in space at the end of the mission.

4. Flyby/Lander Mode

The flyby/lander mission mode was originally suggested by Titus [9]. In this mission the main spacecraft performs a round-trip flyby. An excursion module is detached prior to the Mars flyby to provide the required planetary exploration. The excursion module returns by means of a hyperbolic rendezvous during the planetary flyby of the main spaceship. The trajectory is shown in Figure 1. The major advantage of this mission mode over other stopover missions is the weight saving due to the fact that only the excursion module has to undergo the thrust maneuvers for the landing

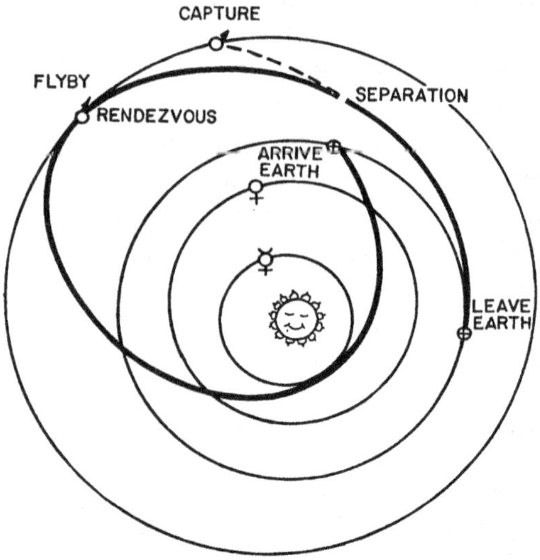

Fig. 1. FLEM trajectory profile [9].

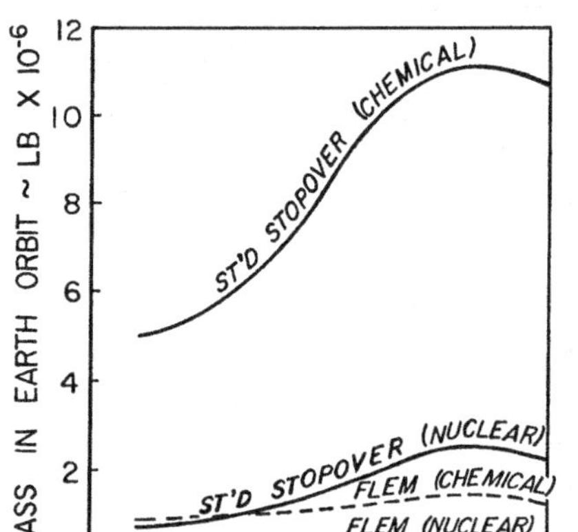

Fig. 2. Mass in earth orbit vs. opposition year [9].

phases. The savings as computed by Titus are shown in Figure 2. Observation of Figure 1 shows that the trajectory goes inside the orbit of Venus and also has a relatively high arrival velocity upon earth return. This observation was exploited by Sohn [10] to achieve additional gains through the inclusion of a Venus flyby to reduce the arrival velocity at earth return. The elimination of retro or aerodynamic capture of the main spacecraft at the destination simplifies the design, relaxes the structural requirements, and eliminates problems associated with parking orbit precession. A failure anywhere in the landing mode does not constitute a complete mission failure. The main spacecraft continues on its own trajectory which returns to earth. The flyby/lander is a logical progression from flyby missions to manned landings. Two major new technical problems appear. One is the planetary entry. The other is the hyperbolic rendezvous of the excursion module. Operational experience through repeated rendezvous with an orbital laboratory would provide confidence for the hyperbolic rendezvous which has but one opportunity for a successful launch.

5. Periodic Orbits

A periodic orbit between the planets consists of a series of free-fall trajectories which allow a spacecraft to sequentially make close encounters with the planets. For the orbit to be periodic, the trajectory after the encounter must allow the spacecraft to free-fall to the near vicinity of another planet. The spacecraft is again accelerated due to the planet's gravity field and continues on to the next planet to be encountered. After a certain period of time, the sequence repeats. Periodic orbits have been shown to exist between earth and Venus [11]. It is possible that periodic orbits exist which visit Mars as well as Venus. That analysis is more complicated and has not yet been performed. The smaller surface gravity of Mars is the complicating factor. Minovitch [12] has demonstrated a multiple flyby which would continue for 5 years making sequential encounters with earth, Venus, Mars, earth, Mars, earth, Venus, earth in that order. Hopefully, longer duration orbits can be found which return to Mars and earth periodically.

Periodic orbits are analyzed by means of patched conics. The trajectory between the planets is an ellipse relative to sun-centered coordinates. The flyby of each planet is a hyperbola relative to planet-centered coordinates. The magnitude of the velocity of the spaceship relative to the planet at the terminal point of the inbound ellipse must be equal to the magnitude of the velocity of the spaceship relative to the planet at the initial point of the outbound ellipse. The magnitude of the relative velocity must be small enough to permit the periapse of the hyperbola to lie above the surface of the planet. To study periodic orbits connecting earth and Venus, the orbital period of Venus is taken to be 8/13 year (an error of less than 0.03% of the observed value). Then the absolute orientations of earth and Venus repeat after 8 years; and an orbit connecting earth and Venus is periodic, in general, if it repeats after a multiple of 8 years. For the special case of circular, coplanar orbits for earth and Venus, the relative orientation of the two planets repeats after only 8/5 years; and a connecting orbit is periodic if it repeats after a multiple of 8/5 years. An orbit which makes 21/5 solar revolutions in 16/5 years is found to satisfy the special conditions if one year is spent near earth making 1 solar revolution and 16/13 years near Venus making 2 solar revolutions. By subtraction the time left for each transfer between earth and Venus is 63/130 year, and the transfer angle is $\frac{3}{5}$ of a revolution. These values allow a solution of Lambert's problem for the orbital elements of the transfer ellipse and the launch and arrival velocity magnitudes. The same transfer ellipse is used going to and from Venus, so the relative velocity magnitudes are identical at the terminal points. The result of the Lambert computation shows the velocity magnitudes to be small enough to permit the flybys to take place above the surface of each planet.

The search procedure in the general case of inclined, elliptic orbits is more complex. The previous orbit now takes 16 years before it repeats exactly, and there are 10 transfers between earth and Venus in the 16 years. Let t_i identify the 10 terminal locations of earth and Venus, and V_i the difference between inbound and outbound relative velocity for the 10 terminal locations. Ten solutions of Lambert's problem are required to find the V_i given the t_i. The problem is to adjust the 10 t_i to make the 10 V_i vanish simultaneously.

The solution has been found by using a steepest-descent procedure. A step is taken in t_i space in the direction which causes the largest change in the sum of V_i^2. The result of the computation shows that the terminal velocity magnitudes are small enough to permit the flyby to take place above the surface of each planet. Other periodic orbits connecting earth and Venus are possible. They are found by substituting alternate trajectories on the earth-to-earth legs and on the Venus-to-Venus legs. About 15 periodic orbits out of a large number of possible combinations have been computed by Menning [13] using procedures similar to those described above. In each case the terminal velocities were found to be small enough to permit the flyby to take place above the surface of each planet. He estimates there are probably over 500 additional satisfactory combinations. The guidance requirements for these orbits have been studied by Hickman [14] and found to be reasonable. The total correction for 16 years was estimated to be only 170 m/sec.

6. Castles in Space

The usual concept of an orbiter or lander mission is one which employs staged rockets. Each stage is abandoned in space after its use and only the last stage returns to earth. A large fraction of the total payload is the main spacecraft which contains the radiation shielding, the shelter, and the life support for the long interplanetary portion of the journey. But this part can be used again. The basic idea here is to construct large, interplanetary spaceships as durable, re-usable, luxurious shelters, literally, castles in space [15]. Under the castles-in-space concept, the main spaceship is left in a periodic, interplanetary orbit, shuttling back and forth between the planets, providing shelter for subsequent missions over several decades. Access to the main spaceship is by a hyperbolic rendezvous during planet flyby. The initial mass of the access vehicle is small because the life support, the radiation protection, and the other massive systems for the long portion of the journey are already in orbit.

A trajectory which shuttles between earth and Venus is presented for the circular, coplanar case to illustrate the concept. A spaceship on this trajectory for 3.2 years would make 4.2 revolutions of the sun while the earth has made 3.2 revolutions and Venus has made 5.2 revolutions. The spaceship stays near earth for one year on an earth-to-earth transfer which makes one revolution of the sun in one year. After a transfer to Venus from earth, the spaceship makes two Venus-to-Venus transfers on orbits which have the same orbital period as Venus, making two revolutions of the sun in 1.24 years. The transfers from earth to Venus and from Venus to earth are on symmetrical portions of identical ellipses located in different parts of the sky. The transfer ellipse is shown in Figure 3. The sketch shows the earth-to-Venus transfer as seen from above the ecliptic plane or the Venus-to-earth transfer as seen from below the ecliptic plane. The two transfers together account for the remaining 1.2 revolutions of the Sun in the remaining 0.96 year. The locations of the terminal planets at each

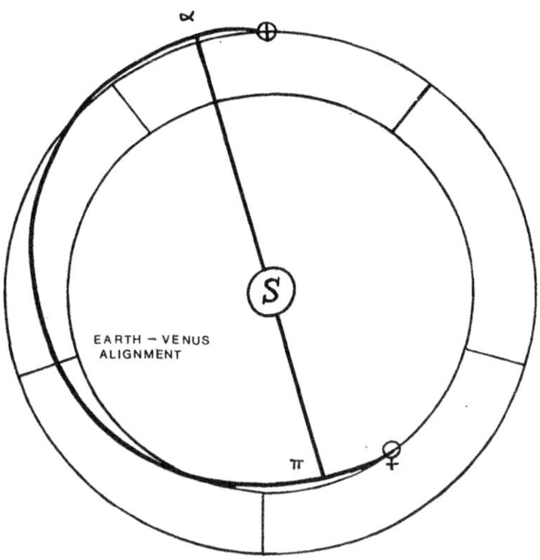

Fig. 3. The earth-Venus transfer. Trajectory data: $a = 0.86$ AU; $e = 0.17$; $t_f = 0.48$ year.

TABLE I

The positions of earth and Venus at the time of flyby

Earth	Venus	Event
0.00	0.00	Reference Alignment
0.18	0.30	I at Venus
0.30	0.49	II at earth
0.80	1.30	I at Venus
1.30	2.12	II at earth
1.42	2.30	I at Venus
1.60	2.60	Alignment
1.78	2.90	II at Venus
1.90	3.09	I at earth
2.40	3.90	II at Venus
2.90	4.71	I at earth
3.02	4.90	II at Venus
3.20	5.20	Alignment

Positions are in revolutions around the Sun starting from the reference alignment. The time of the event in years after the time of the reference alignment is equal to the position of the earth in revolutions from the reference alignment.

flyby define the trip. These locations are given in Table I relative to any alignment of earth and Venus taken as reference for one 3.2 year cycle. It takes five cycles for the trajectory to repeat in the same part of the sky. Extending Table I or Figure 3 to indicate all five cycles would show the terminal points to be equally spaced between the five locations of the earth-Venus alignments. It takes two synodic periods to complete one cycle; therefore, two spacecraft are required to take advantage of every earth-Venus opportunity. These are labeled I and II. When I is on its way to Venus, II is on its way to earth, and vice versa. When I is orbiting the Sun near Venus, II is orbiting the Sun near earth, and vice versa.

Follow the trip by picking up the flight of spacecraft I on its arrival at Venus from earth. During the flyby of Venus the relative velocity vector is rotated out of the orbital plane to put the spacecraft in an inclined orbit with the same period as that of Venus. After one revolution of the sun, the spacecraft returns to Venus. During the second flyby of Venus, the relative velocity vector is rotated again but so as to keep the spaceship on an inclined orbit with the same period as that of Venus. After another rotation of the sun, the spacecraft returns to Venus a third time. During this flyby the relative velocity vector is rotated back into the orbital plane to put the spacecraft on the transfer orbit back to earth. Upon arrival at earth, the relative velocity vector is rotated out of the orbital plane to put the spaceship on an inclined orbit with the same period as that of earth. One year later the spacecraft returns to earth, and the relative velocity vector is rotated back into the orbital plane to put the spacecraft on its trajectory toward Venus.

The trip can be seen in perspective in Figure 4 where the trajectory is shown relative to a coordinate frame that rotates at a rate of one revolution per year. In this frame the earth-Sun line is fixed, and Venus appears to move in a circle, passing earth once every 1.6 years.

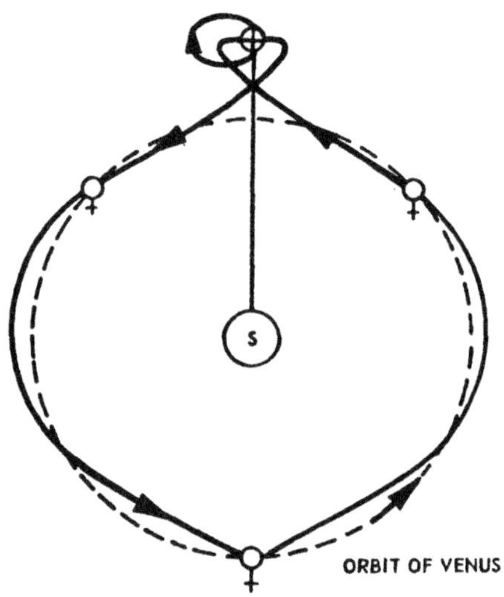

Fig. 4. Periodic orbit I as seen from earth.

With both spaceships operating there is one going to Venus and one returning from Venus at the time of every alignment. During the time spent near Venus, the spaceship goes behind the sun relative to the earth. This would facilitate physical measurements involving the sun by permitting two-way transmission of electromagnetic radiation through the solar corona. The time spent near the earth could be used for replenishment of the spaceship from the earth by unmanned freight ships using continuous thrust. Long transfer times would keep the mass fraction for propulsion of the freightship relatively small. With two spacecraft there are six flybys of Venus and four flybys of earth every 3.2 years. During the course of the trip, the two spacecraft have the opportunity of seeing each planet from all sides. This concept of interplanetary flight is an extension of the flyby/lander concept for advanced interplanetary transportation systems. Most of the operational tasks for the castles-in-space concept would have been demonstrated during the flyby/lander mode. The major new technical problems would be due to the indefinite flight duration, which points up once again the necessity for developing a capacity for sustained operations in space.

7. Operational Need for an Orbital Laboratory

A total, long-range plan for interplanetary exploration would start with the simplest flyby missions, progress toward the flyby/lander mode, and thence to the castles-in-space concept. Each step of such a program would build logically on the accomplishments of the earlier missions. In order to get started, the first step has to be an orbital laboratory because all of the subsequent steps depend upon the operational capabilities which would be developed there.

The most important capability to develop is to be able to carry out sustained operations in space. Safety is of utmost significance for sustained operations. All potential hazards need to be identified and safeguards developed for each. Procedures need to be established for fire, explosion, power failure, contamination, structural failure, etc. The crew needs to be trained adequately to deal with all these emergencies. Regular emergency drills should be conducted and deficiencies corrected. The psychological well-being of the crew is an important safety item. Remedial actions to counter the effects of extended confinement in limited space may have to be taken. Any problem whatsoever that leads to an abort in the orbital laboratory will need to be corrected before a crew is committed to an interplanetary flight. Daily routines need to be worked out. What is to be the length and nature of the work cycle? What does the crew do with its leisure time? How often is it necessary to dump the cabin atmosphere? What standards need to be kept with regard to cleaning? In essence the orbital laboratory provides a 'shakedown cruise' for astronauts and their equipment prior to interplanetary operations. Some sort of repair capability is necessary to meet the high reliability requirement on equipment for interplanetary flight. This needs to be demonstrated in orbit. Tools and spares allowances have to be determined. It must be established what maintenance skills are required among the crew. Confidence in overall equipment reliability must be achieved during actual orbital operations.

Another important operational capability that must be demonstrated for the interplanetary program outlined is a dependable, on-time rendezvous capability. Both the flyby/lander and the castles-in-space concepts utilize a hyperbolic rendezvous. The launch for this rendezvous must be on time because there may be no subsequent launch opportunity or at best a wait of several months. The primary difference between an orbital rendezvous and a hyperbolic rendezvous is the time interval between adjacent launch windows. The larger launch velocity for the hyperbolic rendezvous requires more propellant but does not make the rendezvous any more difficult. The guidance problem for hyperbolic rendezvous may actually be simpler than that for orbital rendezvous. The important consideration is that it be launched on time. Repeated rendezvous with the orbital laboratory will give operational confidence in the ability to perform the hyperbolic rendezvous.

An additional operational utilization for the orbital laboratory is for equipment development, checkout, and testing. Complete operational testing is difficult or impossible in an earth-based laboratory. Such things as the determination of zero-g performance and the testing of new subsystem concepts could be carried out. The orbital laboratory would be performing engineering as well as scientific experiments, i.e., necessary engineering for the next step in the scientific exploration of the solar system.

Finally, the orbital laboratory is needed to find the answer to issues that can only be resolved by experience in orbit. The most important of these issues at this time is the level of artificial gravity required. The weight penalty anticipated is approximately 50% increase in spacecraft weight per g provided. A maximum spin rate of 6 rpm and a minimum acceleration of 0.2 g, for operational effectiveness, was established on the basis of studies in a 1-g environment [2]. The crux of the problem is that experimental data on the exposure of humans to weightlessness is limited to periods of less than two weeks. Longer exposure is required in order to draw a conclusion as to the correct medically and operationally acceptable levels. The only way to achieve longer exposure is in orbit, and the logical means is an orbital laboratory. Another question that can be only answered by experience is the reliability and operational lifetime of spacecraft subsystems. Reliability is critical to the long-range program, and it needs to be proven in space. Finally, if women are to join the crew for the interplanetary missions, the operation should be tried out first in earth orbit. The orbital laboratory would then be performing a sociological experiment valuable in its own right but necessary prior to sending mixed crews to the planets.

8. Summary

The important point made by this paper is that an interplanetary mission requires that a crew be committed to a long duration stay in space. The demand for high reliability during the mission makes it necessary to demonstrate the operational capability for sustained operations in space during Earth orbit where timely abort, escape, and rescue are still feasible. A total, long-range program has been outlined for interplanetary flight. It proceeds from present accomplishments to an orbital laboratory, to simple interplanetary flyby missions, on to the flyby/lander mode of interplanetary exploration, and ultimately to the castles-in-space concept of orbital laboratories on permanent free-fall trajectories between the planets. Each step would build logically on the accomplishments of the earlier missions. The development of an orbital laboratory as a durable, longlife facility is the key link in the overall program leading to manned planetary exploration.

References

[1] Isaacs, T. M. (ed.): *Project NERO*, MIT Press, Cambridge, Mass.

[2] Jenkins, L. E. *et al.*: 1968, 'Spacecraft Requirements for Manned Planetary Missions', AIAA Technology for Manned Planetary Missions Meeting, New Orleans.

[3] Cousteau, J. Y.: 1966, 'Working for Weeks on the Sea Floor', *The Journal of the National Geographic Society* **129**, No. 4.

[4] Hohmann, W.: 1925, *Die Erreichbarkeit der Himmelskörper*, Oldenbourg, Munich.

[5] Crocco, G. A.: 1956, 'One Year Exploration Trip earth-Mars-Venus-earth', in *Proceedings of the XIIth International Astronomical Congress*, Rome.

[6] Battin, R. H.: 1964, *Astronautical Guidance*, McGraw-Hill, New York.

[7] Ross, S.: 1963, 'A Systematic Approach to the Study of Nonstop Interplanetary Round Trips', Interplanetary Missions Conference, 9th Annual AAS Meeting, Los Angeles.

[8] Aeronutronic Division of Ford Motor Company: 1962, 'EMPIRE – A Study of Early Manned Interplanetary Missions', Pub. No. U-1951, Newport Beach, Calif.

[9] Titus, R. R.: 1966, 'FLEM – Flyby-Landing Excursion Mode', AIAA Paper 66-36, AIAA Third Aerospace Sciences Meeting, New York.

[10] Sohn, R. L.: 1967, 'Mars/Venus Flyby Missions with Manned Mars Landers', *Journal of Spacecraft and Rockets* **4**, No. 1.

[11] Hollister, W. M.: 1968, 'Periodic Orbits for Interplanetary Flight', AAS/AIAA Astrodynamics Specialist Conference, Jackson, Wyoming.

[12] Minovitch, M. A.: 1963, 'The Determination and Characteristics of Ballistic Interplanetary Trajectories Under the Influence of Multiple Planetary Attractions', Jet Propulsion Laboratory, California Institute of Technology, Technical Report No. 32-464, Pasadena, Calif.

[13] Menning, M. D.: 1968, 'Free-Fall Periodic Orbits Connecting Earth and Venus', S.M. Thesis, MIT.

[14] Hickman, D. E.: 1968, 'Guidance Requirements for Periodic Orbits', S.M. Thesis, MIT.

[15] Hollister, W. M.: 1968, 'Castles in Space', *Astronaut. Acta* **14**, No. 2.

MANNED FLIGHT TO THE NEARER PLANETS

S. FRED SINGER
Department of the Interior, Washington, D.C., U.S.A.

Abstract. We need a long-range goal for space exploration which is well defined and highly visible. On a 10 year timescale, a manned flyby/orbiter of the planet Mars would give important scientific returns, at a relatively low incremental cost, using building blocks developed in other programs. This project would also lead to major advances in technology, keep open the option of future manned landings, and maintain popular support of the space program. A specific scientific objective of highest interest is the detailed examination of Phobos and Deimos, the natural satellites of Mars. They may be captured asteroids, or they may even be remnant samples of the primeval planetesimals which formed the planets of the solar system.

1. Introduction

The key for maintaining a viable space program, as I see it, is the early definition of a realistic goal for planetary exploration. One needs to set a goal which is technically ambitious, a goal which provides public interest, as well as a stimulus for technical advance. But we must be careful that this goal is not too far out. By this, I mean that it should not be too far away in time nor too expensive, because goals, in order to be achieved, have to be funded.

At the same time, the goal must have some public appeal. It cannot and should not simply be more of the same. It is for this reason that a continuation of unmanned probes over the long term has little appeal to me. But the traditional plan for planetary exploration resembles closely the plan for lunar exploration; namely, a long series of unmanned probes, followed by an orbiter program, followed, in turn, by unmanned landings, and culminating finally in a manned landing.

For this reason, I argue here for an alternative: a manned flyby/orbiter which would, of course, be more difficult, more costly, and later in time than a set of unmanned probes, but less difficult, less costly, and earlier in time than a manned landing. In terms of return of significant scientific data, a manned encounter would probably have a much greater effectiveness than a series of probes of equal cost; and it remains to be shown also that a manned landing can buy for us appreciably more in the way of scientific returns than a manned encounter. In any such cost-effectiveness-analysis, one of the bones of contention will be to define at least the relative value of various scientific data. Let us turn to a discussion of this vexing problem.

2. Scientific Objectives

The general aims of space exploration are agreed upon. They are: to understand the development of the earth; to uncover the origin and development of the solar system; and to trace the existence of life on neighboring planets, especially Mars. But in order to improve cost effectiveness, I would argue that we must be *selective* in targets and

S. F. Singer (ed.), Manned Laboratories in Space.

experiments, concentrating on those that will lead us most directly and most expeditiously to the solution of the major scientific problems. At a time when resources are limited, we must forego a leisurely, gradually-progressing plan of exploration, in favor of a more selective attack which, however, may be further away in time.

To be sure, every bit of information about the solar system has value, but some data are more valuable than others. As I look at the space program, I can see the following happening: The kinds of experiments that were done near the earth and near the Moon tend to be perpetuated near the planets. This is not necessarily the best way to proceed. Just because we measure the atmosphere near the earth doesn't mean that we have to measure the atmosphere near Mars in the same way and with the same precision. Nor does a detailed photographic survey of the Martian surface make sense because we surveyed the moon. Data will be produced and they may be interesting, but they may not illuminate for us the central problems. I am concerned by this phenomenon of persistence, the tendency to continue on with the same types of experiments that have been done before, often just because we have the instruments developed and the techniques of analysis worked out.

One would expect, of course, to find differences of opinion among scientists as to what data are really most crucial. I will simply put forward my own views on what is important.

(1) *Earth problems.* We need to be able to explain the anomalous development of the earth's atmosphere and hydrosphere. Alone in the solar system, it contains oxygen, and alone among the inner planets it contains oceans. It may be the only planet which supports forms of life. It has a magnetic field; it has a large moon. These are important clues which may explain the uniqueness of the earth.

(2) *Planets and satellite systems.* We all want to know how the planets were formed. Why are there such fundamental differences in the terrestrial (inner) planets – Mercury, Venus, Earth, and Mars – and the outer planets – Jupiter, Saturn, Uranus, Neptune? Why are the densities of the inner planets so high? Why don't they have satellite systems? Why does Pluto, the outermost planet, resemble the inner planets?

It is evident that the solar system contains really more than one planetary system, in the sense that Jupiter, Saturn, and Uranus have a system of satellites which mirrors in many respects the planets as satellites of the sun. We may assume, therefore, that the origin of the satellite systems of the outer planets is somehow related to the origin of the planets themselves. But there are certain anomalies which are quite striking. The terrestrial planets do not have satellite systems. Mercury and Venus have no satellites at all. The earth has one satellite, the moon, which is completely anomalous since it has more than 1% of the mass of the earth and five times the angular momentum. It is really more of a sister planet, and some of us are beginning to believe that the moon was captured rather than formed in the vicinity of the earth. The two Martian satellites, Phobos and Deimos, are quite unusual and may hold a clue to the difference between the inner and outer planets' satellite systems.

(3) *Asteroids, meteorites and comets.* We should structure our solar system exploration program to give us as rapidly as possible the answers to the following vital

questions: Was there more than one origin of the solar system, in the sense that 'planetesimals' were formed and reformed, as is suggested by the structure of chondritic meteorites? Are asteroids the fragments of a former planetary body (or bodies) that broke up, or do they represent condensations or agglomerations of planetesimals – opinions are divided. Are meteorites simply asteroids that fell on the earth, or are they remnants of cometary nuclei?

Ideally, we would like to examine some asteroids, some comets, and such bodies as Encke's comet, the Apollo asteroids, Icarus, etc. I particularly want to point to the opportunity of examining Phobos and Deimos; they are so small that they would not have been modified in structure by internal heating and plutonic processes, as, for example, happened with the moon. Also, they are conveniently held in place for us by Mars' gravitational field. They may be captured asteroids or captured cometary bodies; on the other hand, they may have formed out of planetesimals, at the time when the planets formed. This may give us an opportunity to examine directly the original 'stuff' of the solar system. In either case, an examination of their structure and composition would give us important information about the early history of the solar system.

3. Advantages of Manned FlyBy/Orbiter Missions

A number of manned interplanetary mission studies have already concluded that the manned planetary fly-by mission is feasible using Saturn 5-Apollo hardware. Typical mission profiles with low energy injection by modified chemical stage show mission duration of 700 days. A single flight could pass by Mars, pass by Venus, fly close to comets and asteroids, launch probes into the atmospheres of Mars and Venus, even pick up material from the Martian surface and return it, orbit Mars for a length of time, launch probes to Jupiter and beyond.*

Initially, one would have to show the superiority of a manned fly-by over an unmanned fly-by, a superiority which must be large enough to justify the increased costs as well as the delay. I believe that the increased costs can be justified more than adequately in terms of the returns. The delay in the project imposes a scientific penalty which is difficult to estimate.

The advantages of a manned fly-by over unmanned fly-bys are perhaps obvious. The powers of selectivity and judgment of the man have already been mentioned and would, of course, contribute greatly to the success of the exploration. To this should be added the capability of the man to repair instrumentation; the ability of the man to process data immediately; or to recognize a phenomenon as it occurs and focus upon it, followed by making the correct experimental choices as to what to measure and how to measure it.

For an unmanned experiment, you pretty well have to know what you are going to get before you can do it sensibly. In a manned experiment, you can truly explore

* It would be rather an unusual flight which could accomplish all of these missions. The purpose here is to suggest merely what is feasible. Various mission plans have been published from time to time which include all of the activities referred to here. (See also the article by W. M. Hollister, p. 99.)

and our purpose, of course, is exploration. In this connection, I would mention the possibility of furnishing the manned flyby mission with general laboratory instrumentation; e.g., various optical gear, infrared spectrometers, ultraviolet spectrometers, various kinds of particle detectors with various kinds of absorbers. This flexibility would enable the man to put together an optimum experiment for each particular portion of the journey. It would enable him also to take advantage of targets of opportunity, such as meteor streams and comets.

The ability of the man to gather large amounts of data and store them in the spacecraft – or process a certain amount in order to make further experimental choices – is highly important. At the present time, there are severe limitations on the amount of information that can be transmitted back from planetary distances. Because of the time delay, it is virtually impossible to use this information in real time and make experimental choices based upon it. This is a vital point which needs to be stressed again and again. The ability of the man to make an experiment on the spot, based on what he has just observed, gives us a kind of *immediate feedback*, rather than a feedback which extends from one unmanned mission to another, requiring a time interval of perhaps several years.

4. Detailed Experiments

In general, measurements can be done *in situ*, through probes, magnetometers, charged particle detectors, etc., or by remote sensing, using electromagnetic radiation: X-rays, UV, visible (emitted or scattered), infrared (usually emitted), and microwave, either passive or active (radar). Both types of measurements can be improved in performance and flexibility by including a man as the experimenter.

An objection often raised to a manned fly-by is the long flight time, of the order of two years. It is claimed that man would be ineffective and useless for most of this journey. Nothing could be further from the truth. We have to realize that the planets are dynamic and that their atmospheres and surface features are constantly changing. The meteorology of Mars and Venus is an unexplored subject. There will be diurnal and seasonal variations of Mars and Venus which can be observed with different sensitivity from different distances. Observations in the visible range of the spectrum can be carried out as soon as the spacecraft leaves the vicinity of the earth and can be continued over a two-year period. Others, such as infrared observations, may begin when the space probe comes within suitable range and could then extend over a shorter period of time. The experimenter would adjust the optical parameters, such as aperture, wavelength settings, slit width, sensitivity, etc., to optimize the observations.

One of the great advantages of a manned vehicle lies in the possibility of launching instrumented planetary probes, with the possibility of repeating or changing successive experiments based on the results of the previous ones. In case of a failure, the next probe would be suitably adjusted to take care of this error, or to measure in further detail, or with greater sensitivity. This immediate-response capability vastly increases the value of a manned fly-by in terms of data returned. These probes can be launched with very small velocity increments, starting, for example, at distances corresponding to a month or more away from the encounter. Some of these probes could become orbiters. Some would penetrate the atmosphere and measure its composition and that of the clouds, aerosols, and haze. In the case of Mars, it would be quite feasible to construct a probe which lands, takes a sample, and is then picked up by the manned vehicle.

One of the most intriguing possibilities lies in a close approach to Phobos and/or Deimos to observe their shape, structure, albedo, and try to deduce their origin. They may be spherical, if formed by gradual accretion; or they may be odd-shaped fragments from a planetary break-up; or they might just be a loose assembly of rubble held together by weak gravitational forces. We may be able to tie-up to Phobos and take a sample and, from it, determine mineralogical structure and composition, as well as the length of time for which is has been exposed to cosmic rays.

Optical observations of asteroids at closer range would extend our knowledge of their size distribution down to smaller sizes. We could also determine angular velocity, shape, surface albedo, and possibiy composition (by remote sensing).

Measurement of the zodiacal light at different distances from the sun would obtain for us the spatial distribution of interplanetary dust.

Fly-by through a comet tail would allow us to examine its chemical composition and structure, as well as its interaction with the solar wind.

Simple probes could be constructed which would eject smoke particles, or sodium or barium clouds, at various distances from the manned vehicle. By optical tracking of the clouds, one could establish the existence of wind patterns and of electric fields in interplanetary space and unravel the effects of the solar wind on comets. Probes for measuring the solar wind and interplanetary magnetic field could resolve certain ambiguities as to whether changes are due to time variations or spatial variations.

Observations of cosmic rays at different distances from the sun would enable us to investigate interplanetary magnetic field effects on cosmic ray time variations.

There is little question that there is enough to do, to measure, and to analyze, to keep the crew fully occupied.

5. Conclusion

It is clear that over the next five to ten years, planetary exploration can proceed through the occasional launching of simple unmanned probes, similar to Mariner, with the data being used to define an eventual manned landing. On the other hand, we need to keep in mind that with the development of the Apollo and Vostok programs, we will have the capability of an earth-orbiting manned space station, and the capability of testing space life-support systems for periods of many months or even years. The earth space station system also gives us the capability of observing through astronomical telescopes many of the phenomena of the solar system, albeit at low resolution because of the large distance. In a sense, therefore, the space station program could be expanded into the manned fly-by, mainly by adding on propulsion and subsidiary instrumentations.

To summarize, a manned encounter with a planet is what seems to be to me a desirable compromise. It is certainly more costly and further away than a set of simple unmanned probes. It is less costly and closer in time than a manned landing. But, if viewed from a package point of view, in terms of the acquisition of important data on the solar system, then I believe the fly-by alternative is superior.

The course that will be taken by our space program will be conditioned to a large extend by considerations which are not primarily scientific; for example, by the impact of the Apollo Program on the public and on the Congress, and by the financial resources available for space programs generally. While I am in favor of a manned fly-by as a next major objective, I cannot justify its urgency; i.e., that it has to proceed according to an accelerated time schedule. Perhaps over the next ten years, we will have only unmanned planetary probes; I would not consider this a disaster. But if planetary exploration programs were to be dropped completely, this would be a disaster; it would deny us the option to go for a manned planetary program at some later date. It would deny us the opportunity of learning the answers to the most fundamental questions relating to our existence in the solar system.

Bibliography

'The Space Program in the Post-Apollo Period', Report of the President's Science Advisory Committee, The White House, February 1967.

'Space Research–Directions for the Future', National Academy of Sciences, Publication 1403, Washington, D.C., 1966.

'Technology for Manned Planetary Missions', AIAA Meeting, New Orleans, Louisiana, March 1968; proceedings published by American Institute of Astronautics and Aeronautics, New York, 1968.

Manned Planetary FlyBy Missions Based on Saturn/Apollo Systems. Technical Proposal by Lockheed Missiles and Space Company (LMSC-A820555), May 6, 1966.

Experiment Payloads for a Manned Mars FlyBy Mission, Bellcomm Tech. Report 67-233-1, prepared for NASA, May 1967.

ABOUT THE AUTHORS

ROBERT R. GILRUTH is Director of NASA's Manned Spacecraft Center in Houston, Texas. He is a pioneer in manned spaceflight, having been assigned to manage the manned spaceflight program in October 1958. He received his early degrees in aeronautical engineering and served at the Langley Laboratory of the NACA, the forerunner of NASA. He has received an honorary degree from the University of Minnesota, his alma mater. He has been elected a member of the National Academy of Engineering, of the International Academy of Astronautics, and received their Daniel and Florence Guggenheim award.

THEODORE J. GORDON is Senior Research Fellow at the Institute for the Future. Before that, he served for 14 years with the McDonnell-Douglas Astronautics Company in various capacities, including Chief Engineer of the Saturn Program, Test Conductor for the Thor and Thor-derived space vehicles, and finally as Director of Advance Space and Launch Systems. He has served as a Regents Professor at the University of California at Los Angeles, and as a consultant to the Rand Corporation and to various companies. He has authored a number of books dealing with the future impact of technology.

ROBERT L. GERVAIS received a B.S. and M.S. in Aeronautical Engineering with a minor in Nuclear Engineering. During his ten years with Douglas Aircraft Company/ McDonnell Douglas Corporation, he has held various managerial positions in manned spacecraft programs concerned with interplanetary and space station programs and power systems, leading to his present assignment as Project Manager, Space Stations. He is presently a National Director of the American Astronautical Society and Chairman of the Los Angeles Section. He is a charter member of the Systems Engineering Committee of the American Society of Mechanical Engineers.

S. FRED SINGER serves as Deputy Assistant Secretary for Scientific Programs in the Department of the Interior. A geophysicist, he has held university and government positions. While at the University of Maryland, in 1953, he put forward design concepts for a minimum-size instrumented satellite. Ten years later, as the first Director of the National Weather Satellite Center, he was able to adapt these ideas to the design of the operational Tiros satellite. He then served as Dean of the School of Environmental Sciences at the University of Miami, where his research included space oceanography. He was elected a member of the International Academy of Astronautics and has received many awards, including the first Science Award of the British Interplanetary Society, the Gold Medal Award of the Department of

Commerce, and a special citation from President Eisenhower for his work on satellites.

KIRILL Y. KONDRATYEV holds the Chair of Atmospheric Physics at Leningrad University and is currently serving as Rector. He has published numerous papers and monographs in the field of atmospheric and space physics, including one of the earliest books on meteorological satellites. He is Deputy Chairman of the USSR Academy of Sciences' Committee on Exploration and Use of Space and chairs a number of important international committees and working groups. He is a Corresponding Member of the USSR Academy of Sciences and an Honorary Member of the Royal Meteorological Society.

BORIS V. VINOGRADOV is Professor of Geophysics at the University of Leningrad. He has many publications in the field of land surveys, aerial photography, and remote sensing. He heads the Laboratory of Aerial and Space Methods for Earth Sciences at Leningrad University. He is a member of the Committee on Application Satellites of the International Astronautical Federation as well as of other international commissions and working groups.

EDWARD WENK, JR., has served since 1966 under Presidents Johnson and Nixon as Executive Secretary of the Cabinet-level National Council on Marine Resources and Engineering Development. He received his doctorate in engineering from Johns Hopkins University and then served for 14 years in naval research on problems of submarine design. Subsequently, also he served as the U.S. Congress' first Advisor on Science and Technology and founder of their Science Policy Research Division in the Library of Congress. He was recently elected to the National Academy of Engineering and is the recipient of many honors and awards.

WILLIAM T. PECORA is Director of the U.S. Geological Survey in the Department of the Interior. After receiving his Ph.D. in geology from Harvard, he joined the Geological Survey and held various research positions before becoming Director in 1965. He is a member of the U.S. National Academy of Sciences, a foreign member of the Brazilian Academy of Sciences, a member of many advisory committees, and recipient of many honors and awards, among them the Distinguished Service Award of the Department of the Interior. He is a former U.S. intercollegiate fencing champion and represented the U.S. on the Olympic Team in 1936.

LYMAN SPITZER, JR., is Charles A. Young Professor of Astronomy at Princeton University and Director of the Princeton University Observatory. He received his Ph.D. at Yale. At Princeton he has also been in charge of the research on controlled nuclear fusion, and is the originator of the Stellarator Project. Among his many scientific accomplishments, he has given great attention to space research. He is one of the earliest proponents of instrumented satellites and investigated such matters as attitude control, perturbations and instrumentation for satellites. He has been particularly concerned with advancing the concept of a large space telescope. He was elected a member of the International Academy of Astronautics and of the U.S. National Academy of Sciences.

WALTER M. HOLLISTER has been a member of the Massachussetts Institute of Technology faculty since he received his doctor of science degree there in 1963. He is Assistant Professor in the Department of Aeronautics and Astronautics, and a member of the MIT Measurement Systems Laboratory. He has been a guest professor at the Technical University of Berlin and the Swiss Institute of Technology in Zürich.

RÉSUMÉS

R. R. Gilruth: Stations spatiales habitées – porte s'ouvrant sur notre avenir dans l'espace (pp. 1–10).

La communication décrit plusieurs conceptions de stations spatiales habitées, allant des premiers modèles, qui permettront d'évaluer la capacité de l'homme à maîtriser le milieu spatial et à y travailler de façon productive, jusqu'aux bases les plus complexes regroupant une grande variété de satellites scientifiques spécialisés autour d'un noyau central d'hébergement. La communication présente une analyse raisonnée du choix de la taille de la station spatiale, du nombre et des caractéristiques de son équipage, de l'altitude et de l'inclinaison de son orbite ainsi que la durée de son existence et les besoins de réapprovisionnement. La contribution que peuvent apporter les stations spatiales, par le plein emploi des qualités humaines uniques de raisonnement et de travail manuel, à l'observation de la Terre et à l'exploration de l'Univers à l'aide de télescope, est discutée. Parallèlement le développement des stations spatiales et l'expérience acquise au cours des missions opérationnelles, sont envisagés sous l'angle du rôle qu'ils jouent dans la création de la technologie nécessaire pour envoyer l'homme sur les planètes.

T. J. Gordon et R. L. Gervais: Problèmes critiques à l'ingénieur posés par les stations spatiales (pp. 11–32).

Du point d'avancement actuel de nos connaissances, la réalisation de stations spatiales capables de fonctionner pendant longtemps semble faisable et pratique. Chose étonnante, il n'existe que peu de problèmes critiques relevant de la science de l'ingénieur pour lesquels il n'y a pas aujourd'hui de solution unique.

On peut obtenir une longue durée de fonctionnement des sous-systèmes par une combinaison optimale de redondance, de possibilité de maintien et de vérification; l'un des principaux problèmes qui restent à résoudre consiste à déterminer les moyens d'accélérer l'essai de la durée de vie des composants.

Parmi les méthodes dont on dispose pour faire face aux cas d'urgence susceptibles de faire échouer une station spatiale, seuls les modules de libération de l'équipage semblent fournir le court temps de réponse et la grande sécurité nécessaires.

C'est maintenant un fait prouvé que l'homme peut survivre dans une ambiance de pesanteur nulle; cependant, les différentes techniques de thérapeutique psychologique et médicamenteuse dont on a besoin pour assurer l'efficacité de l'homme lors de missions de longue durée restent indéterminées.

Des problèmes de moindre importance comprennent certaines réalisations techniques importantes qui doivent être obtenues pour assurer aux stations spatiales la plus haute efficacité de fonctionnement. Ces questions d'amélioration de l'efficacité comprennent l'optimisation du sous-système et la possibilité de faire concorder stabilité et commande à l'interface, la commande de l'environnement et les systèmes de survie, et les sous-systèmes fournissant de l'énergie.

Tandis que certains problèmes potentiels relevant de la science de l'ingénieur et concernant les stations spatiales de longue durée sont étudiés, l'absence d'un grand nombre de problèmes majeurs est vraiment significative.

S. F. Singer: Satellites météorologiques habités (pp. 33–45).

Les satellites habités peuvent faire progresser la météorologie de diverses façons fort importantes:

(1) – Mettre au point plus rapidement et plus économiquement un système plus fin de capteurs électromagnétiques qui pourrait être employé sur les futurs satellites météorologiques opérationnels;

(2) – Réaliser des programmes d'observation spécialisés, par exemple en relation avec des expériences de modification des perturbations orageuses effectuées à terre;

(3) – Réaliser une fonction d'entretien et de réparation pour les satellites météorologiques avancés dans un système opérationnel futur.

Un expérimentateur humain peut apporter des contributions irremplaçables pour la conception des instruments destinés à la mesure de la distribution des nuages pendant la période nocturne, des vents, des températures, des précipitations, de la teneur en vapeur d'eau et en ozone, des propriétés des nuages, des aérosols, de l'état de la mer, de la neige, de la glace et du terrain.

On met en évidence la valeur économique d'un système de satellites météorologiques d'avant-garde pour ce qui concerne des prévisions météorologiques améliorées; un tel système peut également aider à résoudre les problèmes scientifiques concernant les modifications des climats.

K. Ya. Kondratyev, V. D. Stepanenko, et B. V. Vinogradov: Observation des zones différentes du spectre à partir de satellites (pp. 46–67).

On passe en revue et on discute les possibilités et résultats d'interprétations d'observations de la terre à partir de satellites pour leur utilisation dans les sciences de la Terre et autres investigations des ressources naturelles.

Des zones différentes du spectre électromagnétique peuvent donner des informations différentes et souvent complémentaires sur l'atmosphère, la surface (sol, végétation, océan, glace, hydrologie) et le dessous de la surface.

On examine ici:

(1) La photographie spatiale, en noir et blanc et également multispectrale, et ses applications aux problèmes des ressources naturelles terrestres;

(2) Les images de la Terre en infrarouges dans les fenêtres de transparence et dans les bandes d'absorption, avec les possibilités de leur interprétation météorologique;

(3) La spectrophotométrie spatiale, l'identification des spectres des formations en surface ayant pour but, d'une part, de discerner la composition minéralogique des roches et, d'autre part, des études géobotaniques;

(4) Les 'images radar' de formations en surface, leurs avantages et les possibilités d'analyse active et passive en hyperfréquences.

E. Wenk, Jr.: L'utilisation de véhicules spatiaux pour les études océanographiques (pp. 68–74).

L'utilisation de véhicules spatiaux pour les études océanographiques n'en est qu'à ses balbutiements, mais il est déjà évident qu'elle présente de très grandes possibilités pour l'exploration des océans. Le balayage synoptique à partir de plateformes orbitales peut remplacer le travail d'un grand nombre de dispositifs d'observation à la surface des océans, en ce qu'il permet l'observation à grande échelle des phénomènes océanographiques. Le balayage ou l'exploration à partir de satellites présente des possibilités uniques d'analyses unifiées synoptiques très vastes, voir globales. De telles observations conviennent idéalement pour la décennie internationale d'exploration des océans qui va se dérouler dans les années 1970 à 1980. Cette décennie est envisagée comme devant être un programme de coopération internationale concertée à long terme pour l'exploration des océans à l'échelle mondiale.

Bien que les données océanographiques recueillies à ce jour à partir d'engins spatiaux soient moins quantitatives que les observations à partir de la surface, la décennie à venir verra certainement des améliorations. Combinées avec les données obtenues à partir de navires et de bouées, les observations à partir de véhicules spatiaux offriront un nouveau et efficace moyen pour relever le défi des océans.

W. T. Pecora: Observation des ressources naturelles à partir de satellites (pp. 75–87).

Les photographies obtenues à partir de Gemini et d'autres images réalisées par des satellites ont démontré certains avantages des vues de la Terre à petite échelle prises depuis des satellites. Des satellites à longue durée de vie, fournissant une couverture répétitive uniforme ajouteraient la dimension du temps à notre processus d'interprétation et aideraient à mieux comprendre la dynamique terrestre.

Les géologues, longtemps gênés par le fait qu'ils abordaient les problèmes d'un point de vue nécessairement proche de la Terre et qu'ainsi ils n'étudiaient que des détails pour les extrapoler à l'échelle d'une région ou du globe, attendent de recevoir des images de la Terre depuis l'espace, qui leur permettraient de construire des cartes photographiques des continents et des plateaux environnants. De tels cartes leur permettraient de se concentrer sur les problèmes intéressants et aideraient à la classification des terrains ou à la reconnaissance d'anomalies qui justifieraient des recherches plus poussées. Ainsi les géologues seraient à même, non seulement d'étudier les chaînes de montagne principales et leurs divers éléments structuraux dans leur contexte régional, continental ou à l'échelle du globe, mais aussi viendrait un temps où ils pourraient comparer les variations intervenues à la surface au cours des ans. Les éruptions volcaniques, les tremblements de terre, les incendies de forêts, les inondations et leurs effets sur l'érosion pourraient être surveillés et évalués à partir de l'espace d'une façon rapide et en temps opportun. Cela pourrait conduire vers des améliorations dans les techniques de prévision, d'aide et de rétablissement exigées dans les zones sinistrées.

D'autres phénomènes variables dans le temps, tels les phénomènes de variation du littoral, les variations des estuaires, la formation des deltas et le développement des dunes de sable, ainsi que la croissance ou la destruction des atolls pourraient être enregistrés de façon permanente et on pourrait dresser une carte de leurs modifications. Un grand nombre de ces applications géologiques fourniraient des informations pour d'autres recherches et la planification technique liée à l'utilisation des surfaces, aux centres de peuplement et à la croissance urbaine.

L. Spitzer, Jr.: Recherche astronomique à l'aide d'un grand telescope sur orbite (pp. 88–98).

L'emploi d'un télescope orbital de grand diamètre, avec une ouverture de 3 mètres ou plus, peut revêtir une importance capitale pour résoudre quelques-uns des problèmes les plus importants de l'astronomie moderne. Les caractéristiques de fonctionnement d'un tel télescope sont brièvement passées en revue, de même que les avantages relatifs d'opérations réalisées avec ou sans la présence de l'homme à bord. Quelques-uns des domaines de recherche qui devraient bénéficier particulièrement d'un instrument aussi puissant sont présentés. Ils comprennent en particulier:

(1) Les caractéristiques à grande échelle de l'Univers, sa géométrie, son homogénéité, sa taille et sa courbure.

(2) La structure des galaxies et leur évolution dans le temps.

(3) La distribution de la densité, la composition chimique et l'état physique du gaz interstellaire ainsi que son rôle dans la formation de nouvelles étoiles et dans l'evolution de notre galaxie.

(4) La densité et la température du gaz dans le halo galactique ou la couronne, qui peut provenir de la pénétration de matériel intergalactique ou de l'émission de gaz produite par le noyau de notre propre galaxie.

(5) L'étude suivie et systématique des atmosphères planétaires, y compris les changements de structure en fonction de la position et du temps.

(6) Les détails de la surface de la planète Mercure et son rôle dans les premières étapes de l'histoire du système solaire.

W. M. Hollister: Le laboratoire orbital – un tremplin pour le vol interplanétaire (pp. 99–109).

Un laboratoire orbital est une étape nécessaire dans un programme de vol spatial habité menant à l'exploration planétaire. Des missions vers les planètes demandent un temps de vol total dépassant une année, alors que jusqu'ici les vols habités se sont limités à une durée totale de deux semaines ou moins. Presque tous les problèmes techniques et opérationnels posés par le vol planétaire piloté, indépendamment de ceux posés par le vol lunaire, sont directement liés à ce très grand accroissement de la durée de la mission. Des données expérimentales sur l'exposition d'êtres humains à l'absence de gravité sont nécessaires, pendant ces périodes plus longues, pour déterminer les niveaux de gravité artificielle acceptables du point de vue médical et du point de vue opérationnel pour des vols planétaires. Il est nécessaire d'éprouver les durées de vie en opération des sous-systèmes du véhicule spatial. Pour avoir une grande sécurité il faut une capacité de maintien de vie à bord dont on doit démontrer qu'elle est réalisable en orbite terrestre avant de s'y fier pour un voyage interplanétaire. Le véhicule principal d'une expédition interplanétaire exerce essentiellement les mêmes fonctions qu'un laboratoire orbital. L'un et l'autre jouent le rôle d'abri de longue durée de laboratoire, de dépôt de provision, de station d'arrimage et d'atelier de réparation. En choisissant convenablement la mission interplanétaire, le véhicule principal pourrait en effet être un laboratoire orbital placé sur une trajectoire interplanétaire, qui retournerait sur Terre en chute libre après avoir approché une ou plusieurs planètes. Il est possible de continuer indéfiniment la chute libre, sans autre moyen de propulsion que ce qui est nécessaire à la correction des erreurs de guidage. On esquisse un programme total, à longue terme de vol spatial piloté, qui va des réalisations actuelles à un laboratoire en orbite terrestre, puis aux missions de survol interplanétaire les plus simples en continuant par le mode d'exploration interplanétaire combinant survol et atterrissage, et finalement par les laboratoires orbitaux sur des trajectoires de chute libre entre les planètes. Chaque étape de ce programme se fonderait logiquement sur les réalisations des missions précédentes. Le développement d'un laboratoire orbital en tant que dispositif durable, de longue durée de vie, est le chaînon essentiel du programme général d'exploration planétaire par l'homme.

S. F. Singer: Vol habité orbital des planètes proches (pp. 110–115).

Nous avons besoin d'un objectif à long terme pour l'exploration de l'espace, objectif bien défini et tout à fait visible. Sur une échelle d'une dizaine d'années un vol habité orbital approché de la planète Mars fournirait d'importantes retombées scientifiques et cela pour un prix supplémentaire relativement faible, grâce à l'utilisation de modules contitutifs mis au point pour d'autres programmes. Ce projet conduirait également vers des progrès techniques capitaux, laisserait ouverte l'option pour de futurs atterrissages de véhicules habités et entretiendrait le soutien du public pour les programmes spatiaux.

Un objectif scientifique spécifique du plus haut intérêt est l'examen détaillé de Phobos et Deimos, les satellites naturels de Mars. Ils pourraient être des astéroïdes captés ou même des échantillons, vestiges des corps planétaires primitifs, qui formèrent les planètes du système solaire.

РЕЗЮМА

Р. Р. Гилрут: Пилотируемые космические станции – ворота в будущее космоса (стр. 1–10).

В докладе описываются концепции некоторых пилотируемых космических станций, начиная от самых первых вариантов, которые могли обеспечить основное понимание возможности человека покорить космическое пространство и продуктивно работать в нем, до более сложных космических баз, которые в виде разнообразных специальных научных спутников располагаются вокруг центрального ядра. Рассматриваются принципы, связанные с выбором размера космической станции, количественным составом и характеристикой экипажа, высотой и углом наклона орбиты, продолжительностью функционирования и потребностями снабжения такой станции. Обсуждаются вопросы, связанные с вкладом космических станций в возможность наиболее полного использования способности человека мыслить и действовать при наблюдении Земли и изучать с помощью телескопа Вселенную, а также роль космических станций и опыт их использования для создания необходимой техники для полетов человека на другие планеты.

Т. Й. Гордон и Р. Л. Джервэйз: Важнейшие технические проблемы космических станций (стр. 11–32).

С точки зрения передовых идей нашего времени создание долгодействующих эксплуатационных космических станций представляется возможным и реально осуществимым. Удивительно то, что имеются лишь несколько важнейших технических проблем, для которых в настоящее время не существует какого-либо единого решения.

Долговечность вспомогательных систем может оптимально зависеть от дублирования, эксплуатационной надежности и проверки; основной проблемой остается изыскание средств по ускорению проверки надежности.

Что касается методов, имеющихся для того, чтобы справиться с аварийными неполадками на борту космической станции, только модулы спасения экипажа вероятно дают необходимое время для быстрой реакции и высокую безопасность.

То, что человек может существовать в состоянии невесомости, является теперь неоспоримым фактом; однако еще не определены различные психофипиологлческие, терапевтические и лечебные средства, необходимые для обессечения жизнедеятельности человека во время длительных полетов.

К менее важным проблемам относятся некоторые важные технические доистжения, которые должны быть получены для обеспечения наибольшей эффективности работы космических станций. Эти вопросы по повышению эффективности включают оптимизацию вспомогательных систем и взаимосвязь стабильности и контроля, управления окружающей средой и систем, обеспечивающих жизнь на корабле, а также вспомогательных энергосистем.

В то время, как обсуждаются некоторые потенциальные технические проблемы космических станций большой продолжительности, отсутствие большого числа важных проблем является действительно значительным.

С. Ф. Сингер: Возможности проведения исследований в области метеорологии с помощью пилотируемого космического корабля, находящегося на орбите Земли (стр. 33–45).

Пилотируемые спутники могут содействовать развитию науки об атмосфере в ряде важных направлений. (1) Содействовать более быстрому и экономичному созданию точной системы электромагнетических измерительных инструментов, которая могла бы применяться на будущих метеорологических спутниках. (2) Проводить конкретные программы наблюдений, например, в связи с экспериментами, воспроизводящими условия происходящих на земле ураганов. (3) Осуществлять эксплуатацию и ремонтные операции усовершенствованных погодных спутников в будущих оперативных системах.

Человек-экспериментатор может сделать уникальный вклад в разработку инструментов для измерения распределения облачности в ночное время, ветров, температур, осадков, испарения воды и содержания озона, свойств облаков, аэрозолей, а также состояния моря, снежнего покрова, льда и поверхности Земли.

Подчеркивается экономическое значение усовершенствованных систем погодных спутников для улучшения предсказания погоды; эти системы могут также помочь решению научных проблем, связанных с изменением климата.

К. У. Кондратиев, Д. В. Степанов, и Б. В. Виноградов: Возможности интерпретации изображений Земли, наблюдаемых из космоса в различных областях спектра (стр. 46–67).

К. У. Кондратиев, Д. В. Степанов, и Б. В. Виноградов: Возможности интер-
В настоящем обзоре рассматриваются возможности и результаты интерпретации наблюдений Земли со спутников для использования для геологических наук и других исследований естественных ресурсов.

Различные районы электромагнитного спектра могут давать различную и часто дополнительную информацию относительно атмосферы, поверхности земли (обрабатываемые земли, растительность, океан, лед, гидрология) и подповерхности.

Мы рассматриваем:

(1) космическое фотографирование, черно-белое, а также многоспектральное, для изучения проблем природных ресурсов Земли;

(2) инфракрасные изображения Земли в окнах прозрачности и полосах поглощения и возможности их метеорологической интерпретации;

(3) космическую спектрометрию; идентификацию спекторов земных образований с целью различения минералогического состава горных пород и геоботанических исследований;

(4) 'радиоизображения' земных образований; их преимущества и возможности активного и пассивного микроволнового измерения.

Е. Венк, мл.: Космические аппараты для изучения Океана (стр. 68–74).

Использование космических аппаратов для океанографии находится пока в стадии младенчества, однако имеющиеся в настоящее время признаки свидетельствуют о том, что они имеют огромные возможности для изучения Океана. Синоптическое сканирование с орбитальных платформ может заменить работу многих датчиков, расположенных на поверхности Океана, проводящих наблюдения за крупными океаническими процессами. Сканирование с помощью искусственных спутников Земли открывает неповторимые возможности для проведения широких, даже глобальных, униформированных синоптических анализов. Подобные наблюдения идеально отвечают задачам предстоящего Международного десятилетия изучения Океана, начинающегося в 1970 году. Десятилетие предполагается в виде согласованной, долгосрочной, международной совместной программы изучения Океана в масштабах всей Земли.

Хотя получаемые в настоящее время из космического пространства данные об Океане носят более ограниченный количественный характер, чем наблюдения на его поверхности, в предстоящем десятилетии несомненно произойдут улучшения. В сочетании с данными, получаемыми с помощью судов и буев, наблюдения с космических аппаратов явятся новым и полезным средством познания океанов.

В. Т. Пекора: Изучение ресурсов Земли с орбитального космического корабля (стр. 75–87).

Фотографии с 'Джемини' и других космических аппаратов продемонстрировали известные преимущества мелкомасштабных изображений Земли из космоса. Находящиеся длительное время в полете спутники, дающие униформированное повторяющееся изображение, добавляют временный процесс к нашему процессу познания и содействуют лучшему пониманию динамики Земли.

Геологи, длительное время ограниченные подходом, связанным с изучением деталей Земли, и экстраполяцией их на региональные и глобальные масштабы, готовятся к получению изображений Земли из космоса, которые позволят им сконструировать фотокарты континентов и окружающих их шельфов. Подобные карты позволят сосредоточить внимание на интересующих их проблемах и помогут провести классификацию территорий или известных аномалий, которые нуждаются в дальнейшем изучении. Геологи получат возможность не только исследовать основные горные цепи и их различные структурные элементы в их региональном, континентальном или глобальном контексте, но они также смогут своевременно провести сравнение изменений поверхности, которые происходят на протяжении ряда лет. Можно будет быстро и своевременно регистрировать и давать оценку извержений вулканов, землетрясений, лесных пожаров, наводнений и их последствий. Это позволит улучшить предсказания, оказание помощи и проведение восстановительных работ, которые требуются в районах бедствий.

Можно будет непрерывно регистрировать и наносить на карты другие изменяющиеся во времени явления такие, как процессы изменения побережий, эстуариев, образование дельт, изменения дюн, рост или разрушение атоллов. Многие из этих геологических применений предоставят информацию для будущего планирования исследований и строительных работ, связанных с использованием земель, населенными центрами и ростом городов.

Л. Спитцер. мл.: Астрономические исследования с помощью крупного орбитального телескопа (стр. 88–98).

Исследования с помощью крупно-размерного орбитального телескопа с апретурой в 120 дюймов или больше могли бы представить большой интерес для решения некоторых наиболее важных проблем современной астрономии. Приводятся краткие рабочие характеристики подобного телескопа, а также соображения относительно сравнительных преимуществ его работы с помощью и без помощи людей. Рассматриваются некоторые области исследований, которые могли бы особенно выиграть от использования такого мощного инструмента, в том числе:

(1) крупномасштабные характеристики Вселенной, включая ее геометрию, однородность, размеры и искривления;

(2) стурктура галактик и их изменение во времени;

(3) частота распределения, химический состав и физическое состояние межзвездного газа и роль этого газа в образовании новых звезд и в эволюции нашей галактики;

(4) плотность и температура газа в галактическом гало или короне, которые могут возникать из выпадающего межгалактического вещества или из газа, выбрасываемого ядрами нашей собственной галактики;

(5) непрерывное и систематическое изучение планетарных атмосфер, в том числе изменения структуры в соотношении с положением и и временем;

(6) детали поверхности планеты Меркурий и их влияние на раннюю историю солнечной системы.

У. Холлистер: Оръитальная лаборатория – шаг к межпланетным полетам (стр. 99–109).

Орбитральная лаборатория является необходимым звеном в программах пилотируемых космических полетов по исследованию планет. Для полетов на планеты требуется время более одного года, в то время как сейчас время пилотируемых полетов не может превышать двух недель. Почти все технические и оперативные проблемы, связанные с межпланетными пилотируемыми перелетами, превышающими по длительности и расстояниям полеты на Луну, непосредственно связаны со значительным увеличением продолжительности миссии. Для таких продолжительных периодов требуются экспериментальные данные о пребывании человека в условиях невесомости с целью определения допустимых с медицинской и оперативной точек зрения уровней искусственного притяжения для межпланетных полетов. Необходимо показать сроки работы субсистем космического корабля. Для высокой надежности требуются бортовые технические характеристики, которые должны быть предварительно испытаны на земной орбите до их обеспечения для межпланетных полетов. Главный космический корабль межпланетной экспедиции выполняет в основном те же функции, что и орбитальная лаборатория. Каждый служит в качестве убежища, лаборатории, склада, посадочной станции и ремонтной мастерской. Благодаря правильному выбору межпланетной миссии главный космический корабль может фактически быть орбитальной лабораторией, выведенной на межпланетную траекторию, предусматривающую свободное возвращение на Землю после облета одной или нескольких планет. Возможно продолжать межпланетный свободный полет бесконечно, используя двигатели лишь для исправления навигационных ошибок. Дается описание общей долгосрочной программы пилотируемых космических полетов от современных достижений к лаборатории, помещенной на земную орбиту, затем к простейшим межпланетным облетным миссиям, далее к межпланетным экспедициям, предусматривающим облет и посадку, и, наконец, к орбитальным лабораториям, выведенным на свободную траекторию между планетами. Каждый шаг этой программы будет логически основываться на предшествующих достижениях. Создание орбитальной лаборатории в качестве надежной установки, рассчитанной на длительное использование, является ключевым звеном общей программы, ведущей к пилотируемым миссиям по исследованию планет.

С. Ф. Сингер: Пилотируемые полеты к ближайшим планетам (стр. 110–115).

Нам необходима четко разработанная и ясно видимая долгосрочная задача для изучения космического пространства. В ближайшие 10 лет пилотируемый полет около планеты Марс или орбитальный ее облет мог бы дать важные научные данные при сравнительно небольших расходах, если использовать конструктивные элементы, разработываемые для других программ. Этот проект мог бы вызвать также к жизни важные технические достижения, открыть возможности для будущих высадок космонавтов и содействовать оказанию широкой поддержки общественностью программе космических исследований. Наиболее интересная конкретная научная задача заключается в детальном изучении естественных спутников Марса – Фобоса и Деймоса. Они могут предстдвлять собой попавшие в зону притяжения астероиды или даже остатки первичного планетезималя, из которого образовались планеты солнечной системы.

INDEX OF SUBJECTS

GPSR Compliance

The European Union's (EU) General Product Safety Regulation (GPSR) is a set of rules that requires consumer products to be safe and our obligations to ensure this.

If you have any concerns about our products, you can contact us on ProductSafety@springernature.com

In case Publisher is established outside the EU, the EU authorized representative is:

Springer Nature Customer Service Center GmbH
Europaplatz 3
69115 Heidelberg, Germany

Batch number: 10614940

Printed by Printforce, the Netherlands